Delphin RASOLONJATOVO

Population, distribution and ecology of the Madagascar fish eagle

Delphin RASOLONJATOVO

Population, distribution and ecology of the Madagascar fish eagle

ScienciaScripts

Imprint

Cover image: www.ingimage.com

This book is a translation from the original published under ISBN 978-620-6-70991-6.

Publisher:
Sciencia Scripts
is a trademark of
Dodo Books Indian Ocean Ltd. and OmniScriptum S.R.L publishing group

120 High Road, East Finchley, London, N2 9ED, United Kingdom
Str. Armeneasca 28/1, office 1, Chisinau MD-2012, Republic of Moldova, Europe
Managing Directors: Ieva Konstantinova, Victoria Ursu
info@omniscriptum.com

Printed at: see last page
ISBN: 978-620-8-37040-4

Contents

GLOSSARY

- **Protected Area:** a clearly defined geographical area that is recognised, dedicated and managed by legal means. legal or other means to promote the conservation à long-term of nature and ecosystem services andes tural values(IUCN).
- **Use rights:** the removal of natural resources for non-commercial purposes to satisfy the domestic, vital or customary needs of the resident local population. They are non-transferable and are exercised within the framework of the Community Management Agreement.
- **Fady**: that which is sacred, forbidden, prohibited, abominable, incestuous, that from which one abstains or from which one must abstain, cursed.
- **Management of a Protected Area:** management of all the actions to be carried out in a Protected Area, the aim of which is to enable it to fulfil its ecological, economic and social functions in a sustainable manner.
- **Protected Area Manager:** any public or private person, mixed group, legally constituted group or local community managing the Protected Area in collaboration with the stakeholders concerned.
- **Loa-drano:** sacrifice offered to the deity led by the "Tompon-drano" (literally Master of Water) to mark the opening of the fishing season.
- **Network: a** set of Protected Areas linked by common objectives, common management principles, a common manager or common interests.
- **Philopatry**: the tendency of certains individuals to stay or instinctively return to the place (on land or underwater) where they were born, to reproduce.
- **Polyandry**: refers to themating system by which thefemale of a speciesanimal species mates with severals maless over the course of a breeding season.
- **Ramsar site**: area, "with water that is static or flowing, fresh, brackish or salt, including areas of seawater at a depth not exceeding six metres at low tide", conserved and rationally used through local, regional and national actions and international cooperation (article 1.1 of the Convention).
- **Wetlands**: defined as land, whether worked or not, that is usually permanently or temporarily flooded or saturated with fresh, salt or brackish water. These areas are transitional spaces between land and water (ecotones). Like all these types of special areas, they have a high biological potential (specific flora and fauna) and play a role in regulating run-off and improving water quality (Convention on Wetlands, 1998).

INTRODUCTION

Located in the east of the African continent, Madagascar is home to 23 species of birds of prey, 17 of which are diurnal and 6 nocturnal (Collar *et al.*, 1994; Fuchs *et al.*, 2007). Of these 23 species, 3 of the 12 endemic species are classified among the rarest birds in the world, namely the Red Owl *Tyto soumangnei*, the Madagascar Serpent Eagle *Eutriorchis astur* and the Madagascar Fish Eagle *Haliaeetus vociferoides* (Collar *et al.*, 1985; Collar *et al.*, 1994).

The Madagascar fish eagle is the island's largest endemic bird of prey (Watson, 1993). It is the only critically endangered species of diurnal bird of prey on the African continent (Meyburg, 1986). In the past, this species was not rare; it was observed all along the Malagasy coast and was common in the north-western part (Oberlé, 1981). However, its status changed from "endangered" (EN) to "critically endangered" (CR) between 1987 and 2010 (Birdlife International, 2010). In 2006, its remaining population was 287 individuals. Its range was limited between the district of Morombe, in the south, the bay of Befotaka, in the north, and 120 km inland from the coast, (Razafimanjato, 2011).

The Madagascar fish eagle is mainly found in wetlands. However, because of their resources, these areas are the most affected by anthropogenic pressures due to the incessant needs of the human population linked to demographic growth. They have now become a focus of interest for local residents and fishermen (Ratsisompatrarivo *et al.,* 2016). The report on the state of the environment in Madagascar (MEEF, 2012) also indicates that, throughout the country, wetlands are also affected by water variations, sometimes severe as a result of the intensification of cyclones, but above all by sedimentation, silting and filling. *Haliaeetus vociferoides* is threatened by the loss of its natural habitat as a result of the over-exploitation of lakes and rivers, the conversion of wetlands into rice fields, deforestation and the direct hunting of adults, the collection of eggs and eaglets from nests and captive breeding. Added to this are increasingly intense tropical cyclones (Watson, 1993; Ralison *et al.*, 2011; Razafimanjato, 2011).

Faced with this critical situation for the Madagascar fish eagle, "The Peregrine Fund" project, which has been working in Madagascar since 1992, has taken the initiative to safeguard this species in their preferred areas. These initiatives include raising awareness, increasing the size of the population through semi-captive breeding, involving local residents in GELOSE, providing support in the Tsimembo Manambolomaty and Mandrozo Complex Protected Areas (PAs), and various studies. These include, among others reproductive biology and home range (Razafindramanana, 1995; Rafanomezantsoa,1998), habitat requirements and hunting behaviour (Berkelman, 1997), sex and dominance taking into account paternity and group structure with two male individuals (Tingay, 2000), bio-ecological and impacts of the application of "GELOSE" (Razafimanjato, 2011) all in the Antsalova district and the global census in 2005 and 2006.

Despite this, only a few of the sites that are home to this species are protected by their Protected Area status, and the species is still not a conservation target for the management areas. Individuals living outside these PAs are still susceptible to all sorts of pressures. What's more, the current status of its population is less well known, as the last global census was carried out a decade ago. It is with this in mind that this study focuses on **"a study of the population size, distribution and ecology of the Madagascar fish eagle *Haliaeetus vociferoides,* in the western part of Madagascar".** Its overall aim is to assess the population and current state of conservation of the Madagascar fish eagle in order to improve its protection through the involvement of the managers of the study sites.

To achieve this overall objective, specific objectives have been set for this study:

- inventory the population in each type of ecosystem (coastal, lake and river) and in the various study sites;
- compare the distribution of staff between protected and unprotected areas;
- revise the geographical distribution;
- analyse the population growth or decline over the last 10 years (2006 - 2016);
- break down the variation in the population and reproductive success in the two TPF intervention

sites (Antsalova and Maintirano) over this decade;

- to characterise the nest, the nest tree and the habitats surrounding the nests of this species; - to collect data on the pressures and threats weighing on the species and its habitat.

Based on these objectives, four working hypotheses were put forward:

- 1: the distribution of *Haliaeetus vociferoides* between protected and unprotected areas is homogeneous;
- 2: the limits of the geographical range of this species remain unchanged over this decade;
- 3: the population of this species is growing;
- 4: nest height and Dhp of nest tree are the same between continental and coastal environments.

This document will first describe the methods used, followed by the results and their interpretation, as well as the related discussions. Finally, the conclusions and recommendations relevant to the study are given.

CHAPTER I

I. STUDY ENVIRONMENTS

This study was carried out at 11 sites: Aire Marine Protégée (AMP) Nosy Hara, Aire Protégée Marine et Côtière (APMC) Sahamalaza - ìles Radama, Lac Tseny, Pare National (PN) Ankarafantsika, PN Baie de Baly, District de Besalampy, District de Maintirano, District d'Antsalova, District de Belo-sur-Tsirbihina, District de Miandrivazo, and AP Complexe Mangoky - Ihotry (CMI) (Figure 1).

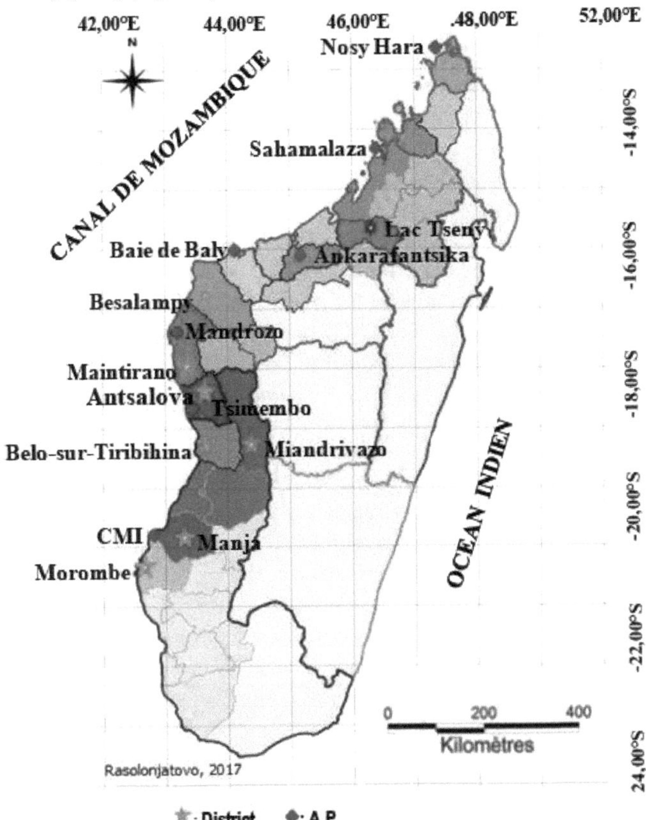

Figure 1: Map of the study area (Rasolonjatovo, 2018).

1.1.Nosy Hara Marine Protected Area

The Nosy Hara MPA is located in the District of Antsiranana II, DIANA Region, between the geographical coordinates 12°06' - 12°28'S and 48°39 - 49°03'E. From north to south, it straddles four (4) rural communes, including Andranofanjava, Andranovondronina, Mahalina and Mangaoka. The total surface area of this PA is 125,471 ha. It is made up of three parcels subdivided into three zones: hard core (32,310 ha) and buffer zone (93,161 ha), including a service zone of 49 ha (Ralison *et al.,* 2011). The habitat type is highly varied: karst and basalt islands, sandy banks or banks sometimes associated with coral debris, sandy or muddy littoral zones and beaches, open sea, mangrove forests, dense dry forest and bushes (TPF, 2013a).

The Nosy Hara MPA is known for the presence of the smallest chameleon *Brookesia micra* (Figure 2), a reptile measuring 2 to 3 cm in length (Glaw *et al.*, 2012).

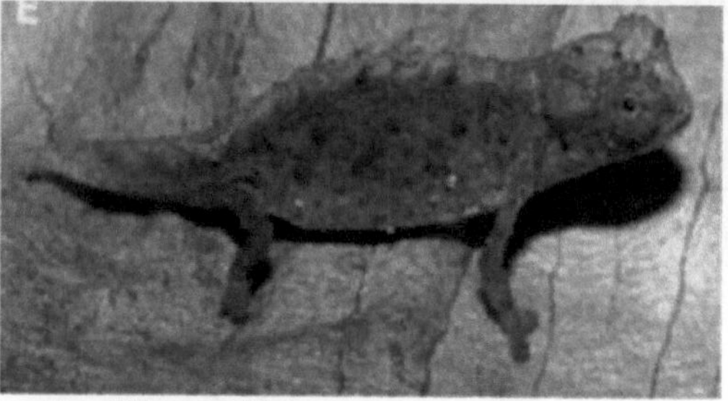

Figure 2: *Brookesia micra* (Glaw, 2012).

1.2.AP Sahamalaza - Radama Islands

Sahamalaza - Radama Islands is a Marine and Coastal Protected Area (MCPA), located in the north-west of Madagascar, straddling the Analalava District (SOFIA Region) and the Ambanja District (DIANA Region). Sahamalaza Bay is a marine and coastal transition zone but also a major estuary located between the geographical coordinates 13°52' - 14°27'S and 47°36' - 47°46'E. With a surface area of 26,035 ha, this AP spreads over five rural communes, namely, Ambolobozo, Ankaramy, Befotaka, Maromandia (Analalava District) and Anorotsanga (Ambanja District) (Clauvice, 2014). The Sahamalaza CMPA is made up of four habitat types such as dry forest, wooded savannah hills, coral reefs and mangroves. The latter covers around 10,000 ha in the bay made up of five parcels including Ampasimbezo, Ankitsika, Kapany, Maromandia and Sijoro.

The climate is warm sub-humid. The average annual rainfall is 1,600 mm and the average annual temperature varies between 26 and 27°C. Rainfall occurs throughout the year, especially from October to April, with a maximum in January (ZICOMA, 1999). The Aire Protected Coastal Marine Sahamalaza-iles Radama is known for the presence of a species of amphibian *Boophis tsilomaro* that is locally endemic and classified as "critically endangered" (Andreone *et al.,* 2010).

Figure 3: *Boophis tsilomaro* (Andreone, 2010).

1.3.Lake Tseny

Located in the rural Commune of Tsaratanana, District of Port-Bergé, Sofia Region,

Mahajanga Province, between the geographical coordinates 15°40'00 "S and 47°51'00 "E, Lake Tseny is one of the priority sites for biodiversity conservation in Madagascar listed in the Ecosystem Profile of the MAD-IO Hotspot. The lake is currently managed by the Madagasikara Voakajy Association (MAVOA) and the Association Privée pour la Production d'Alevins d'Andapa (APPA). It is a key site for the conservation of Madagascar's endemic fish, particularly *Paretroplus menarambo* (Damba menarambo), a tropical fish of the family Chichlidae family. In 2004, *Paretroplus menarambo* was classified as Extinct. In 2010, it was rediscovered in Lake Tseny (Andriafidison *et al.*, 2011) and classified as "critically endangered" (CR) in the IUCN Red List (2017).

1.4.Ankarafantsika National Park

Ankarafantsika National Park is located in the north-west of Madagascar, in the District of Marovoay, Boeny Region. It has a total surface area of 130,026 ha. Crossed by national road n°4 linking Antananarivo to Mahajanga, the Park is 115 km from the town of Mahajanga and 454 km from Antananarivo. It is bounded to the east by the Mahajamba river and to the west by the Betsiboka river. Ankarafantsika National Park plays an important economic role at local and regional level. It is a reservoir of water for the surrounding plains and especially for the rice-growing plain of Marovoay, Madagascar's second largest rice granary. Its forest is the last remnant of dense dry deciduous forest in the north-west, a centre of endemic fauna and flora (Koechlin & Guillaumet, 1974).

1.5.Baly Bay National Park

The Baie de Baly was classified as a National Park by decree no. 97-1452 of 18 December 1997 (ZICOMA, 1999). This park is located in the District of Soalala, Boeny Region, precisely between the geographical coordinates 15°57' - 16°08'S and 45°17' - 45°22'E. It covers an area of 69,350 ha. Its altitudinal variation is fixed between sea level and 100 m. Baie de Baly National Park is the only site in the world that is home to the endangered turtle species *Astrochelys yniphora* (Figure 4), known by the Malagasy name "Angonoka". It was specifically created to safeguard this species (MNP, 2016).

The climate is warm sub-humid, with a marked dry season lasting up to 7 months (April to October). The average annual temperature is around 27.8°C. However, the hottest month is March, with an average temperature of 28°C, compared with 24.2°C in July, considered the coldest month. The Baie de Baly region receives 1,774 mm of rainfall a year, with 5 months without a drop of rain from May to September (ANGAP, 2002). The hydrographic network includes numerous seasonal streams and two large permanent rivers, the Andranomavo and the Kapiloza. The site contains several permanent lakes, of which Lake Sarka (around 300 ha) is the most dominant (Razafimanjato, 2011).

Figure 4: *Astrochelys yniphora* or "Angonoka" (Internet 1).

1.6.Besalampy district

Besalampy is the largest district in the Melaky Region, covering an area of 1,129,200 ha. The climate is tropical, hot and dry, with two distinct seasons: the dry season, from April to October, and the rainy season, from November to March. The highest temperatures correspond to the most intense rainy period, which is favourable for annual rainfed crops (ZICOMA, 1999). This area belongs to the warm sub-humid tropical zone at low and medium altitude, with annual rainfall varying between 600 mm and 1,200 mm. The Besalampy wetland complex is a combination of coastal habitats, rivers, lakes and their forests (Razafimanjato, 2011). It is a constant source of water for rice-growing areas during the dry season. The availability of permanent water is likely to boost numerous economic activities such as river and sea transport, fishing and agriculture. The continental wetlands of Besalampy, such as lakes Amparihy, Ampandra and Sahapy, are well-known for the presence of endangered bird species such as *Haliaeetus vociferoides, Amaurornis olivieri (*Olive Rail), *Ardeola idae (*White Crab Heron)*, Ardea humbloti* (Madagascar Heron*), Anas bernieri* (Bernier's Teal) and a species of freshwater turtle endemic to Madagascar, *Erymnochelys madagascariensis* or "Rere" (Figure 5) (Rabenandrasana *et al.*, 2009).

Figure 5: *Erymnochelys madagascariensis* (Velosoa, 2008).

1.7.Maintirano District

The Maintirano district is geographically located in the mid-west of the island, in the Melaky region. It covers an area of 45,825 ha. The climate is warm sub-humid, with two distinct seasons: a dry season from April to October, and a wet season from November to March. It is a tropical region with an annual rainfall of around 1,525 mm, peaking in January. The average annual temperature recorded in 2015 was 27.17°C (Pruvot, 2016). The wetlands in this district are made up of rivers, coastlines, lakes and swamps. Two long rivers, Manambao and Ranobe, in addition to permanent lakes, are a constant source of water for rice-growing areas during the dry season in Tambohorano (Oberlé, 1981).

For this study, the part north of the Manambao River that encompasses the Tambohorano wetlands, i.e. the Mandrozo PA, and the surrounding wetlands were the subject of a census of the Madagascar fish eagle.

With a surface area of 1,800 ha, Lake Mandrozo is the fifth largest lake in Madagascar (TPF, 2013b). It lies 60 km north of Maintirano, precisely between latitudes 17°31' - 17°33'S and longitudes 44°02' - 44°06'E. This lake is known to be important for at least 2 endangered bird species, including *Haliaeetus vociferoides* and *Amaurornis olivieri* (Figure 6) (Razafimanjato, 2011; Pruvot, 2016). It has been an integral part of the ZICOMA since 1999 (ZICOMA, 1999) and was classified as a "Ramsar Site" in 2012 (TPF, 2013b). Lake Mandrozo also provides a habitat for ichthyofauna, creating a source of income for the local population through fishing activities.

Figure 6: *Amaurornis olivieri* in Lake Mandrozo (Pruvot, 2016).

1.8.Antsalova district

The district of Antsalova is located in the Melaky region and lies between the geographical coordinates 18°40'S - 44°37'E. The Bemaraha Plateau, the Mozambique Channel, the Manambolo River and the commune of Soahany form its eastern, western, southern and northern boundaries respectively. Its total administrative area is 697,119 ha. The region has a dry tropical climate, with a marked contrast between a dry season from April to October and a very hot rainy season from November to March (Rabearivony *et al.*, 2010). In 2015, rainfall was 2,071 mm, with the most recorded during the first quarter. Temperature extremes occurred in July (with an average of 21.3°C) and December (with an average of 27.8°C). The average annual temperature for this year is 25.04°C (Raveloson, 2017).

This area is one of the places in Madagascar to have obtained various qualifications, both national and international. The national Site of Biological and Ecological Interest (SIBE) for the Manambolomaty lake/forest complex in 1988, World Heritage status for the Tsingy in 1990, Ramsar status for the Manambolomaty complex in 1999 and the PA for the Tsimembo Manambolomaty Complex in 2015 are concrete examples. The Antsalova area is rich in biodiversity, particularly aquatic birdlife. Six species of waterbird in the area are threatened with extinction: *Haliaeetus vociferoides*, *Ardea humbloti*, *Anas bernieri*, *Threskiornis bernieri* (Sacred Ibis), *Charadrius thoracicus* (Madagascar Gravelot) and *Ardeola idae*

(Young, 1995; Birdlife International, 2010; Raveloson, 2017).

1.9.Belo-sur-Tsiribihina district

The district of Belo-sur-Tsiribihina belongs administratively to the Menabe Region. It covers an area of 801,699 ha. The climate is warm sub-humid, with an average monthly temperature of between 19°C and 36°C. Rainfall is low and varies greatly from year to year, with annual averages of 700 to 1,000 mm. The rainy season extends from January to March (ZICOMA, 1999). The upper Tsiribihina wetland, at an altitude of no more than 200 m, encompasses the Tsiribihina river, several lakes and rivers, as well as lakes between the Miandrivazo district and the Malaimbandy rural district (Razafimajato, 2011). In the Tsiribihina valley, between Miandrivazo and Malaimbandy, there are a number of lakes where the incursion of the water hyacinth *Eichhornia crassipes* is remarkable and a large expanse of marshland dominated by *Cyperus* sp and *Phragmites* sp. These lakes, which are fairly shallow, are connected to each other during the rainy season.

1.10. Miandrivazo District

The wetland of the Miandrivazo district, in conjunction with that of Malaimbandy, is an integral part of the upper Tsiribihina. This district is upstream and to the east of Belo-sur-Tsiribihina. As a result, the Tsiribihina river flows through the Miandrivazo district, passing through the Belo-sur-Tsiribihina district and emptying into the Mozambique Channel.

1.11. AP Complexe Mangoky - Ihotry (CMI)

The Mangoky - Ihotry Wetland Complex is located in the Atsimo Andrefana and Menabe Regions, Morombe and Manja Districts, on the south-west coast of Madagascar, at latitude 21°25'22.14"S and longitude 43°25'44.13"E. It is a protected area decreed in 2015 (Decree no. 2015-719 of 26 February 2015), covering a total surface area of 426,146 ha. It is bounded to the north by the Commune rurale
d'Andranopasy (District of Manja), to the South by the rural Communes of Befandefa and Basibasy (District of Morombe), to the East by the rural Commune of Nosy Ambositra (District of Morombe) and to the West by the Mozambique Channel (Durrell, 2014). The complex is made up of dense dry forests, thorn forests (Humbrert & Leandri 1954), lakes including Lake Ihotry and the surrounding seasonal lakes (South), mangrove swamp comprising the Mangoky River delta including the mudflat area and maritime space (North) and the swamps around the Ankiliabo road and Nosy Ambositra (the far East). The CMI PA has been identified as an area of importance for bird conservation in Madagascar since 1999 (ZICOMA, 1999). It is known for the presence of endangered bird species including *Threskiornis bernieri, Ardea humbloti, Ardeola idae and Anas bernieri*, and also has significant economic importance in terms of fishing and agriculture. Every year, the Mangoky river overflows its banks and regularly renews the fertility of the rice fields and other lowlands (baiboho). This area is characterised by the arid to hot semi-arid climate of western Madagascar. Annual rainfall is around 600 mm. The rainy season extends from November to March, while the dry season runs from April to October. The average annual temperature is over 20°C. The absolute minimum and maximum recorded are 42°C and 5°C respectively (Rakoto, 2005).

Table 1: Distribution of study sites according to ecosystem type.

Sites	Types of ecosystems		
	Coastline	Lake	River
A.P Nosy Hara	Yes	No	No

A.P Sahamalaza - Radama Islands	Yes	No	No
A.P Ankarafantsika	No	Yes	No
A.P Baie de Baly	Yes	Yes	No
Lake Tseny	Name	Yes	No
Besalampy district	Yes	Yes	No
Maintirano District	Yes	Yes	No
Antsalova district	Yes	Yes	Yes
Belo-sur-Tsiribihina district	Yes	Yes	Yes
Miandrivazo District	No	Yes	Yes
CMI	Yes	Yes	Yes

CHAPTER II

II. MATERIALS AND METHODS

II.1 SPECIES STUDIED

II.1.1. Systematics

The Madagascar fish eagle is the subject of this study. It belongs to the following classification:

Kingdom Animal
Phylum : Vertebrates
Class : Birds
Order : Accipitriformes (AOU, 2010)
Family: Accipitridae
Genus : *Haliaeetus*
Species : *vociferoides* (DesMurs , 1845)

Common names:

In Malagasy: *Ankoay, Vorobe, Vorohôlo*

In French: Aigle pêcheur de Madagasca ou Pygargue de Madagascar

In English: Madagascar Fish Eagle

The Accipitriformes group together diurnal birds of prey such as eagles, vultures, buzzards and hawks. For a long time, these bird species were classified with falcons in the order Falconiformes. But recent studies have led ornithologists to separate them into two very distinct orders. These species are characterised by their short, hooked beaks and powerful, clawed legs called talons. They are usually excellent daytime hunters with keen eyesight. Some, like vultures, feed on carrion. They nest in trees or on rocks and cliffs. The larger species generally raise just one or two chicks a year. The young are nesters.

The Accipitridae, a cosmopolitan family, includes the majority of diurnal birds of prey. They vary considerably in size, shape, plumage, habits and habitat (Langrand, 1995).

II.1.2. Morphological description

The Madagascar fish eagle is one of two species of eagle found on the island. It has a wingspan of around 2 m and a body length of 70 to 80 cm (Langrand, 1995). The adult male (Figure 8) weighs between 2,150 and 2,350 g, or 2,215 g on average (n = 3 individuals), compared with 2,750 to 3,000 g (2,915 g on average, n = 3 individuals) for the female (Figure 7) (Razafindramanana, 1995). Like other raptor species, *Haliaeetus vociferoides* therefore displays sexual dimorphism in terms of size, with the female being larger than the male (Newton, 1979; Langrand, 1995; Razafindramanana, 1995; Tingay, 2000). Adults are characterised by brown plumage with a white tail, head and neck ribs. Their legs are yellowish white. However, in juveniles (Figure 9), the lower part of the body and the wings are marked with buff at the apex with a tawny brown breast (Morris & Hawkins, 1998).

Figure 7: Female *Haliaeetus vociferoides* (Rasolonjatovo, 2018).
Figure 8: Male *Haliaeetus vociferoides* (Rasolonjatovo, 2018).

Figure 9: Juvenile *Haliaeetus vociferoides* (Rene de Roland, 2017).

11.1.3. Lifestyle

The Madagascar eagle is a diurnal species, inactive for most of the day. This bird often stays on its lookout post or perches at a certain high vantage point. It can often be seen alone or in pairs high up. It watches for its prey at this height, mainly fish on the surface of the water. It launches itself and catches these fish using its clawed feet. Endowed with the ability to fly, it can travel considerable distances. As they travel, they soar high into the air, describing a spiral-shaped trajectory. During these flights, it is usually molested by other birds. When the attack is intense, it reacts by turning completely on its back and abruptly thrusting its feet backwards with extended talons. It often perches high on a tree at the water's edge. The eagle adopts a sedentary lifestyle but the young disperse to breeding areas (Langrand, 1987; Rafanomezantsoa, 1998).

11.1.4. Habitat

The Madagascar fish eagle lives in a variety of ecosystems such as lakes and rivers with adjacent forests, and also mangroves, islets and rocky coastlines (Langrand, 1987; Langrand & Meyburg, 1989;). In the western part of the island, it is most often found on rivers and lakes, where it generally depends on large trees on the banks and nesting perches (Rabarisoa *et al.*, 1997). It prefers deep, clear lakes (Berkelman *et al.*, 1999; Watson *et al.*, 2000). In the north-west, it frequents marine islets and mangrove-lined bays along the coast where the water is calm and shallow, with its many fish (Berkleman *et al.*, 2002; Razafimanjato, 2011).

Figure 10: Habitat types of the Madagascar fish eagle (Rasolonjatovo, 2018);

(a): Lake Soamalipo and the surrounding forest, in the Antsalova district,
(b): mangrove of APMC Sahamalaza - Radama Islands, Ampasimbezo plot,
(c): rocky marine island in the Nosy Hara MPA.

11.1.5. Diet

The Madagascar fish eagle is a fish-eating bird. Its diet is based on fish from the CHICHLIDAE family. This family is represented by three species, *Oreochromis macrochir* (Voaloky*), O. mossambicus* (Mahaimiteraky) and *Tilapia zilii* (Borivava), but it also feeds on other fish such as *Arius madagascariensis (Gogo), Glossogosius giuris (Banàna), Megalops cyprinoides (Fiafoty) and Ophicephalus striatus (Fibata*) (Razafimanjato, 2011). The choice of fish depends on its relative abundance. However, in the coastal areas of north-west Madagascar, the Fish Eagle feeds on fish such as *Fistularia spp*, *Tylosurus spp* and in rare cases attacks young domestic ducks and waterbirds such as *Anhinga melanogaster*, *Phalacrocorax africanus*, *Ardea cinerea*, *A. humbloti*, *Dendrocygna viduata* (Berkelman *et al.*, 1999). Eagles feeding on *Platalea alba,* crabs and young sea turtles can also be observed (Rabarisoa, 2000).

11.1.6. 1.6. Reproduction

Under favourable conditions, a pair of Madagascar fish eagles can reproduce every year. A female may mate with two males at the same time, giving rise to what is known as "polyandry" (Tingay, 2000; Tingay *et al.,* 2002). The breeding season begins in May with nest building. Each pair chooses its own nesting site. The egg-laying period occurs between the last week of May and the third week of July. The number of eggs per brood varies from one to three. The time lag between the first and second clutch is two to four days. The third clutch is reported only once, three days after the second (Razafimanjato, 2011). Incubation begins as soon as the first egg is laid. It lasts between 41 and 44 days after the first egg is laid. In general, each member of the pair, whether a pair of two or three individuals, takes part in brooding the eggs, but with a predominance of females (Watson *et al.*, 1999; Razafimanjato, 2011). The hatching period for this species is between the second week of July and September. After hatching, the growth period of the chicks varies from 78 to 89 days and the chicks generally move from the nest to other nearby branches around day 70ème (Watson *et al.*, 2000). At the age of 14 to 28 weeks, the juveniles are able to fly and their parents begin to abandon them gradually. The young Fish Eagle becomes an adult and acquires its definitive plumage after its 4ème or 5ème year (Langrand & Meyburg, 1984; Rafanomezantsoa, 1998) and its longevity is estimated at over 30 years in the wild.

11.2. EQUIPMENT USED IN THE FIELD AND ITS USES

The following equipment was used to carry out this study:

- A pair of binoculars (10 - 40X ZOOM) and a telescope: for observing the individuals and the contents of the nest (egg or chick);
- A GPS (Global Positioning System) device such as the Garmin Etrex 10: for take the geographical coordinates;
- A Nikon Coolpix digital camera, (24.7 megapixels): for shooting illustrative and reference photos;
- A watch with a stopwatch: this allows you to measure the time taken for activities;
- Nylon rope and a tape measure: to measure the height of the nest;
- Geographical maps: for fine-tuning sites;
- Climbing accessory (rope and jumar): for direct observation inside the nest;
- Notepad, pens, pencil: to record the data needed in the field;
- Motorbike, speedboat or pirogue: to get around the sites and to travel from one site to another.

11.3. DATA COLLECTION METHODS

11.3.1. Field study period

This study took place from July 2016 to January 2017, a period of 7 months. During this period, field visits were carried out in 3 stages. In the mid-western part of the island, i.e. from Besalampy to Belo-sur-Tsiribihana, the study was carried out from 03 July to 26 September 2016. In the north-western part, the survey was carried out between 08 October and 11 November 2016. And the south-western part of the island, from Miandrivazo to Morombe, was visited between 12 December 2016 and 13 January 2017.

11.3.2. Census technique

The census mainly concerns individuals of the species studied and their nests. Two complementary methods were adopted:

- The survey;

- Direct observation.

11.3.2.1. Survey

The aim of this method is to identify all the places likely to be home to the Madagascar fish eagle and to understand the situation of the species under study in the face of the various pressures and threats exerted on this species and its habitat. To achieve these objectives, individual and/or group surveys were carried out at each study site.

11.3.2.1.1. Individual survey

In the mid-western and south-western parts of the island, the survey was conducted entirely on an individual basis, as it was difficult to find local people willing to give up their time for a meeting. In total, two hundred and thirty-eight (238) people were surveyed. The majority were male (around 80%), as most of the women refused to answer the questions and gave way to their husbands. In addition, many are reluctant and give vague or false answers without exact results.

The types of people surveyed are made up:

- local authorities: mayors or deputy mayors of municipalities, local councils, etc.

neighbourhood chiefs, the chiefs of the fishermen's camps in the areas visited;

- members of Vondron'Olona Ifotony (VOI);
- Olobe" elderly people;
- farmers or peasants and fishermen (fish or crabs) camping out

around the lake or mangrove swamp that is the target of the visit;

- PA technicians;
- the Mangoky-Ihotry Complex site manager and site agent;
- the chief forester of the Besalampy District.

During the conversation, we first explain our mission and then briefly describe the morphology of the Madagascar fish eagle by showing photos. We then provide information on the presence or absence of places likely to be home to this species (lakes, rivers or mangroves), and finally look at the different kinds of threats and pressures weighing on the species and its natural habitat.

11.3.2.1.2. Group survey

In the four protected areas of Madagascar National Parks (MNP) such as Nosy Hara, Sahamalaza, Ankarafantsika and Baie de Baly, the surveys were carried out in groups rather than individually. For the group survey, the directors of these PAs were asked to group their teams together. As a result, the Nosy Hara and Sahamalaza teams were grouped together in the Nosy Hara MPA office. The group was made up of a sector manager and a CLP for AMP Nosy Hara, and a sector manager and three CLPs for APMC Sahamalaza. In the Ankarafantsika and Baie de Baly PAs, the surveys were carried out by the TPF team. During the meeting, we:

- explained the description of the osprey: its morphology, its range, its habitat

distribution, its current status and its reproductive biology;

- assessed their knowledge of the presence of the Madagascar fish eagle in

specifying the number and location of individuals, nests and chicks for each territory, both inside and on the outskirts of the PAs. The results for each site were tabulated.

11.3.2.2. Direct observation

After the surveys, field visits were carried out. All the activities were based on the collection of data on the species, in particular individuals and nests with their contents in eggs or chicks, and studies relating to certain nesting parameters. For these, systematic direct observation was the method adopted. This method consists of exploring places likely to be home to this species. The observers were equipped with a number of items suitable for this type of work, such as maps, GPS, binoculars and a telescope.

11.3.2.2.1. Searching for individuals and nests

The search for nests and individuals differs between coastal areas and continental wetlands. During the survey, the means of travel also vary according to accessibility (Appendix 1).

> ***Coastal ecosystem***

In all coastal areas, the surveys were based mainly on information received during the surveys. Visits to the north-western part were carried out with PA and/or CLP officers, while the mid-western and south-western parts were surveyed with TPF technicians and fishermen available as local guides. However, the search was not limited to the locations identified during the surveys, but also included other areas likely to harbour this species. Observations were made using binoculars and a telescope, at a distance of around 100 m from the coast. Counting began at around 6 a.m. when the tide was high, stopped when the tide was low and resumed when the tide was high (Razafimanjato *et al.*, 2014).

In this type of ecosystem, the search for individuals and nests varies according to the type of habitat.

- **Islands**

All the rocky marine islets such as Nosy Hara, Lakandava, Anjombavola and Ankerambalavo, the mangrove and dry forest marine islets such as Nosy Valiha and Mahanonoka, were visited all around until individuals and nests were discovered.

- **Mangroves**

In the north-western part, with the exception of the Ankitsika-APMC Sahamalaza plot, the entire mangrove formation both inside and on the periphery of the protected areas was surveyed using speedboats.

In the mid-western and south-western coastal zone of the island, the search was carried out at the same time as that of the continental wetlands. The research was carried out from Besalampy in the north to Morombe in the south. The visits were mainly confined to the mangroves of the following localities: Sonenga, Besalampy district; Tambohorano, Maintirano district, Masoarivo, Antsalova district, and Andranopasy, Manja district, as well as in the Tsiribihina and Belo-sur-Mer estuaries.

> ***Inland wetlands***

In continental wetlands, a total of 64 sites were visited, including 58 on lakes and 6 on rivers (Appendix I). The survey was carried out on motorbikes, then on foot or by pirogue, depending on the circumstances. In areas where dugout canoes were available, the survey was carried out by criss-crossing the coast at a distance of around 100 m from the shore. This distance may vary depending on the height of the canopy of plant cover at the lake or river bank. If this is not the case, the observer will cross the forest on the edge of the lake while looking at the large trees likely to be the perch or nesting place for this species. Whatever the means used, the speed of movement during the observations did not exceed 5 km/h and the activities always began around six o'clock in the morning. All the individuals seen and/or heard were therefore counted.

II.3.2.2.2. Observation of nest contents

As soon as the nest was found, the supporting tree was climbed using the equipment provided (rope and jumar). All the contents (eggs or chicks) were then counted and photographed. Sometimes, all the observer had to do was stand at a fairly high level, around 30 m from the tree where the nest was located. For best visibility, this distance may vary depending on the topography of the area and the height of the nest tree. A pair of binoculars was used to see inside the nest.

11.3.3. Study of nest and nest tree characteristics

The shape and location of raptor nests vary widely (Village, 1990). A number of parameters have been determined for this purpose, namely :

- the condition of the nest: whether it is new or old, and whether it is intact or demolished;
- shape,
- size, the measured parameters of which are illustrated in Figure 11,
- the number of branches supporting it.

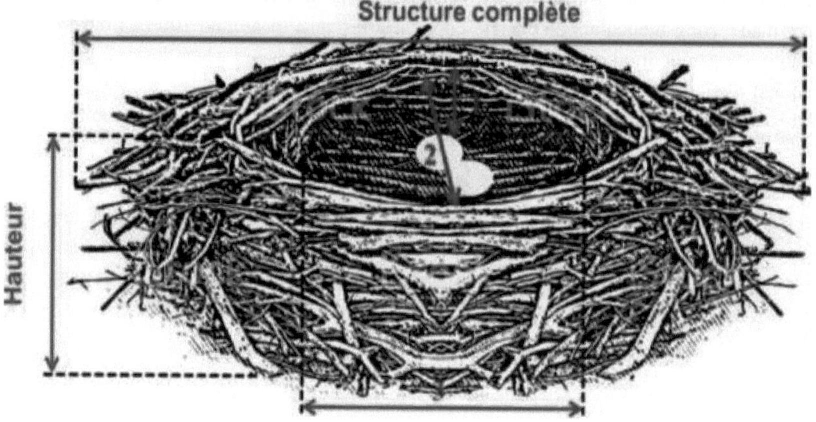

1: Depth, 2: Internal width, 3: Internal length

Figure 11: Illustrative diagram of an eagle nest (iStock, 2017).

The height of the nest from the ground was measured using a rope stretched to the ground and this length was marked with a knot. This rope was then measured using a tape measure.

In order to complete the information on the nest, samples of the constituent branches were measured (lengths and diameters).

The choice of nest tree is important for each pair. The strategy that the species uses during reproduction can be known from this choice. To characterise the nest tree, the following parameters were collected:

- its Dhp or diameter at chest height;
- its vernacular name.

The names of nesting trees were determined with the help of local guides familiar with western forest trees. They were then converted into scientific names using the "Dictionnaire des noms malgaches des végétaux" established by Boiteau (1999).

In general, birds of prey are loyal to their territory. The factor triggering this nest change has therefore been identified.

11.3.4. Study of the habitat surrounding the nest

The following parameters were recorded for each nest found:

- the geographical coordinates of the nesting territory;
- the distance between two neighbouring nest trees of *Haliaeetus vociferoides;*
- the distance of this tree from the nests of other raptor species;
- the distance of the nest from man-made human disturbance factors (camp or village, cultivated land, etc.) within a radius of 2 km of the nest;
- the distance of the nest from a permanent body of water in continental environments;
- the characteristics of the nesting territory.

This method allows us to learn about the interrelationship of the species studied with the environment and also with other raptor species.

11.3.5. Assessment of threats and pressures in the habitat complex

During the visit to each site, all kinds of anthropogenic threats to the habitat and their impact on the way of life of *Haliaeetus vociferoides* were studied. Findings were made by direct observation of the littoral zone, at lake level and within a 50 m radius of the nesting sites. Certain parameters were used to assess the degree of threats and pressures weighing on the species and its habitat (Raselimanana *et al.*, 2012). These parameters are as follows: the collection of eggs, the removal of nesting eaglets and hunting, the number of large trees cut, taking into account their distance from nesting trees, land clearing and bush fires. In the case of lakes, the presence of overfishing (intensive fishing even during the resource renewal period) and the encroachment of floating vegetation, as well as the use of banks as pasture or their transformation into rice fields, are among the data collected.

In order to gather information, surveys were carried out among the local population and protected site staff.

II.4. DATA ANALYSIS

This section illustrates the data analysis methodology. The data were used to verify the 4 hypotheses set out in this study. The data was manipulated and formatted using Microsoft Excel. Mapping was carried out using ArcGIS 9.3, and statistical tests were performed using XLSTAT, version 2016.

11.4.1. Species distribution

The nature (homogeneous or heterogeneous) of the distribution of numbers between protected and unprotected areas was determined using the Chi-square test of homogeneity. To do this, a number of sites were grouped together in order to obtain theoretical numbers greater than or equal to five (5).

All the GPS coordinates of the locations hosting the target species obtained during the censuses were used to produce the updated map of the geographical distribution of *Haliaeetus vociferoides*. In areas not included in the initial list of study sites, we took into account information obtained from personal communications with other researchers and protected area managers.

11.4.2. Population trends

This involves determining the direction or orientation of population change, i.e. whether the population of the species in a well-defined location is increasing or decreasing over time (Andreianov *et al.*, 2011). Evolution was assessed by the rate of change in population size between two dates compared with its starting value. This method has been applied to demography, but this time it was adopted to assess the evolution of the Madagascar fish eagle population in the sites included in the list of target sites for the 2005 - 2006 census. The

results of this census are reported in Razafimanjato's thesis (2011).
The rate of change "t" is calculated using the following formula:

$$t = \frac{(E_1 - E_0)}{E_0}$$

E0: Initial workforce E_1 : Current workforce

Here, the term "evolve" means to increase or decrease.
If t > 0, the population increases over time.
If t < 0, the population is decreasing.
If t = 0, the population size remains constant over time.
. If the population is growing, the average annual growth rate R_a is calculated using the following formula (Velpry, 2015):

$R_a = (E1/Eo)^{1/N} - 1$ or $R_a = V(E1/Eo) - 1$ where

N = number of years between E_1 and E $_{.0}$

This average annual growth rate R_a is expressed as a percentage.

11.4.3. Breeding success at the 2 TPF sites and surrounding areas

In the district of Antsalova (noted: site 1) and in the wetlands of Tambohoano and its surroundings (noted: site 2), a Madagascar Fish Eagle nest is visited three (3) times a year. The first visit coincides with the egg-laying period, while the second and third visits correspond respectively to the periods when the eggs hatch and the chicks fledge. Visits are not limited to the Tsimembo Manambolomaty and Mandrozo PA complexes. They are also made to all the satellite lakes and surrounding bodies of water. As a result, all the data from 2006 to 2016 are already available at the TPF (Appendix 7).
Breeding success is broadly defined as the fledging of chicks. It is calculated from the ratio between the number of successful fledged nests and the total number of active nests.
Note that a nest is said to be active if it contains one or more eggs for a given biological year.

11.4.4. Statistical analysis

> **Mean and standard deviation**

Certain results, such as the distances between two neighbouring nests, the distances of the nests from the water and others, are represented in the form of the mean with its standard deviation (mean ± standard deviation) according to the following formulae:

$$\overline{X} = \frac{\sum_{i=1}^{n} x_i}{n} \qquad s = \sqrt{\frac{\sum_{i=1}^{n}(x_i - \overline{X})^2}{n-1}}$$

Where
X: mean of variable *x* of rank n *S*: standard deviation
x_i: Value of variable *x* at rank*n:* number of samples

> **Chi-square test**

The Chi-squared test is used to check the homogeneity of the facts observed with those to be expected for two or more samples (Johnson, 1992). This test is the most widely used and recommended for ornithologists (Fowler & Cohen, 1985). In this study, it was used to compare the distribution of population numbers in protected and unprotected areas. To this end, the three sites of Lac Tseny, Ankarafantsika and Baie de Baly were grouped together. The Besalampy and Maintirano sites were merged and the samples from Belo-sur-Tsiribihina and Miandrivazo were pooled in order to obtain theoretical values greater than or equal to 5. The following assumptions were made:

H_0: "The distribution of population numbers in protected and unprotected areas is homogeneous".

H_1 : "The distribution of population numbers in protected and unprotected areas is heterogeneous".

The Chi-square value is obtained using the following formula:

$\chi^2 = \sum_{i=1}^{k} \frac{(O_i - C_i)^2}{C_i}$ Where O_i: observed value. C_i: Theoretical value and

$$C_i = \frac{(\text{nombre total de ligne}) \times (\text{nombre total de colonne})}{\text{la somme totale}}$$

$$\frac{(\text{total number of rows}) \times (\text{total number of columns})}{\text{the total sum}}$$

The degree of freedom is calculated as follows: **ddl = (l - 1) (c - 1)** where: **l** = number of rows, **c** = number of columns.

In the case where the degree of freedom is equal to one (**d.d.l.=1**), we have to go through the Yate correction (Fowler & Cohen, 1985) with the following formula:

$$\chi^2_{yate} = \sum_{i=1}^{k} \frac{(|o_i - c_i| - 0.5)^2}{c_i}$$

The value of / at final will be 2 x χ^ .αiβ

ηη

To evaluate this test, the observed Chi-square value (/ cal) is compared with the critical value (/ tab), taking into account the ddl for a p threshold at $\alpha = 0.05$.

If $\chi^2_{cal} > \chi^2_{tab}$, the difference is significant and the null hypothesis H_0 is rejected.

22

If $X_{cal} < X_{tab}$, the difference is not significant and Ho is accepted.

> **Shapiro-Wilk test**

The Shapiro-Wilk test is a statistical analysis designed to examine the compatibility of an empirical distribution with the normal distribution (Rakotomalala, 2011). It is also known as a normal distribution goodness-of-fit test. Compared with other tests, it is particularly powerful for small samples ($n < 50$).

The test statistic is :

$$W = \frac{\left[\sum_{i=1}^{\frac{n}{2}} a_i \left(x_{(n-i+1)} - x_i\right)\right]^2}{\sum_i (x_i - \bar{x})^2}$$

With:

- X¿: corresponds to the sorted data series;

- n: sample size;

- a^: constants generated from the mean and the variance-covariance matrix of the quantiles of a sample of size n following the normal distribution.

The higher *W* is, the more credible the compatibility with the normal distribution.

In this study, this test was used to determine the nature of the distribution of the data for a significance level of 0.05 with a null hypothesis stipulating a normal distribution of the data.

- If during the test the probability $p > 0.05$, the distribution of the data is normal. In this case, we then perform a parametric test for statistical analysis.

If, on the other hand, the probability $p < 0.05$, the distribution of the data is abnormal. This suggests using a parametric nominal test to analyse the data.

> **Student's t test**

Student's t-test is a test of the difference between means used to compare two means. More specifically, it is a parametric test that measures the statistical significance of the difference between two means (Léo *et al.*, 2012). For the present study, this test is used to determine whether nest height varies between continental environments (lake and river) and the littoral zone.

As hypotheses:

- **H0**: "There is no difference in the average nest height between the different environments.

continental and coastal".

- H1: "The average nest heights between these two environments are different".

The value of "**t**" is obtained by the following formula (Fowler & Cohen, 1985):

$$t = \frac{\overline{X_1} - \overline{X_2}}{Se\sqrt{\frac{1}{n_1} + \frac{1}{n_2}}}$$	$\bar{X}$ = average ***Se*** = common standard deviation ***n*** = number of employees

The significance level associated with the test is 0.05.

If the calculated probability is less than 0.05, the null hypothesis is rejected and the difference is statistically significant.

If, on the other hand, the calculated probability is greater than 0.05, the difference is not significant.

This means that we accept the null hypothesis.

> **Wilcoxon test**

The Wilcoxon Test is a non-parametric test used to determine the value of differences (Di = after - before) for the same sample. In the present study, this test was used to determine whether the variation in the population size of *Haliaeetus vociferoides* between 2006 and 2016 was significant. For this purpose, only the sites in the present study were considered, with the exception of Lake Tseny. In addition, we wish to test the null hypothesis according to which the difference in population numbers resulting from the two censuses (in 2005 - 2006 and in 2016) is zero. The alternative hypothesis states that this difference is not zero.

The significance level associated with the test is 5%. If the calculated probability is less than 0.05, the null hypothesis is rejected, i.e. the variation in population size is significant.

If, on the other hand, the calculated probability is greater than 0.05, the difference is not significant. In this case, the population does not show an expressive evolution during this decade.

- **Mann-Whitney test**

The Mann-Whitney test (Whitney & Mann, 1947) is a non-parametric alternative to the t-test for comparing the means of two independent samples from the same population (Fowler & Cohen, 1985; Johnson, 1992). For this study, the comparison of the means of Dhp of nesting trees between the two environments (continental and coastal) requires the use of this non-parametric test since the distribution of the data does not follow the normal distribution. The null hypothesis (**H0**) was: "The difference between the two means of Dhp is equal to 0".

For the same reason of choice, this same test is also used to determine whether or not there is a significant difference between two sites, Antsalova and Tambohorano Wetlands, during the last ten years of monitoring eagle reproduction in Madagascar. The null hypothesis is: "Each stage of reproduction (laying, hatching and fledging) remains the same at these two sites over the decade".

The process:

Each observation value was assigned a rank and the U test was calculated using the following formula:

$$U_1 = n_1 n_2 + \frac{n1(n1+1)}{2} - R1$$

$$U_2 = n_1 n_2 + \frac{n2(n2+1)}{2} - R2$$

Where **n1** and **n2** represent the number of people in each sample.

R1 and **R2** represent the rank numbers.

To complete this test, continue with the following steps:

- put all the data together in increasing order;
- calculate the sum of each row **R1** and **R2**;
- calculate the **U1** and **U2** tests;
- check that **n1.n2** is equal to **U1+U2**;
- check the value of **U** on the table, taking **n1** to always represent the smallest value of two sites;
- determine the value of Z if the number of samples (n) is greater than 20 with.

$$Z = \frac{U - \frac{(n1.n2)}{2}}{\sqrt{\frac{n1.n2(n1+n2+1)}{12}}}$$

If the value of **U** on the table (at the critical threshold of $p = 0.05$) is greater than the calculated value, the null hypothesis *H0* is rejected and the difference is significant. On the other hand, if the value of **U** on the table is lower than the calculated value, the difference is not significant. Otherwise, if the value *ofZ* **e]-1.96; +1.96[,** the test is not significant. If Ze]-**1.96; +1.96[**, the difference is significant and the null hypothesis **H0** is rejected.

Among **U1** and **U2**, choose the smallest value, which will be the value of U.

CHAPTER III

III. RESULTS AND INTERPRETATION

111.1.SURVEYS

The individual survey enabled us to note the various places likely to host the Madagascar fish eagle, while the group survey provided information on the presence of the species in the four target PAs.

111.1.1. Individual survey

From the Besalampy district to the Miandrivazo district, 56 locations were recorded according to 179 people surveyed. These 56 sites are divided into three types of ecosystem: 49 lacustrine, two coastal and five riverine (Table 2).

Tableau 2: *Distribution of Fish Eagle habitats reported during individual surveys, according to ecosystem type from Besalampy to Miandrivazo.*

Sites	Types of ecosystems		
	Lake	**Coastline**	**River**
Besalampy	4	1	-
Maintirano	6	-	-
Antsalova	22	-	3
Belo-sur-Tsiribihina	8	1	1
Miandrivazo	9		1
Total	**49**	**2**	**5**

However, none of the 59 people surveyed confirmed the presence of this species between Belo-sur-mer and Morombe. Those responsible for the Mangoky Ihotry Complex site, which includes the Manja and Morombe district wetlands, have also confirmed the absence of the Madagascar fish eagle in this area. Their last academic monitoring in 2011, the annual monitoring in 2014 and the community monitoring in 2015 all reported the absence of individuals of this species. The last sighting of a pair of osprey in Lake Bemamba, Fokontany de Vatolampy, Manja district, was in 2007.

2.1.2. 2. Group survey

The atmosphere in the group was friendly, the respondents were active and the discussion was interesting.

At Nosy Hara, 16 territories, 31 adult individuals, eight (8) nests and one chick were recorded. These 31 adult individuals are divided into 15 normal pairs (n = 30), i.e. a pair composed of two individuals, one male and one female, and one solitary. Of these 16 territories, 10 and six (6) are located inside and outside the PA respectively (Table 3).

Tableau 3: *Results of a group survey on the presence of the Madagascar fish eagle in Nosy Hara.*

Location	Territory	Adult individual	Nest	Chick
In the PA	Nosy Nosy Hara	2	2	1
	Nosy Lakandava	2	1	-
	Ankonkofaly	2	1	-
	Andovobe	2	1	-
	Fararano	2	1	-
	Nosy Anjombavola	2	1	-

	Andranomavo	1	-	-
	Rantavorona	2	-	-
	Antsakoa	2	-	-
	Andranoboka	2	-	-
On the outskirts of the PA	Nosy Tanga	2	-	-
	Nosy Sahankazo	2	-	-
	Nosy Valiha	2	1	-
	Nosy Mahanonoka	2	-	-
	Nosy Androtso	2	-	-
	Ampasindava	2	-	-
Total	**16**	**31**	**8**	**1**

At Sahamalaza - Radama Islands, 23 individuals including 19 adults composed of nine (9) normal pairs (n = 18 individuals) and one solitary individual, and four chicks were recorded. All the territories (n = 10) are located within the National Park (Table 4). No nests were reported during the survey.

Tableau 4: *Results of a group survey on the presence of the Madagascar fish eagle in the Sahamalaza - Radama Islands site.*

Location	**Parcel**	**Territory name**	**Adult individual**	**Nest**	**Chick**
In the PA	Kapany	Anj iamangintagna	2	-	1
		Anjiamalandy	2	-	-
		V avanankaramihely	2	-	1
		Ankerambalavo	2	-	1
	Ampasimbezo	Ankij animanj ofo	2	-	-
		Ankingagnivo	2	-	-
	Sijoro	Vavanambodidimaka	2	-	1
		Nosy Kalanoro	2	-	-
	Maromandia	Ankorobe	2	-	-
	Ankitsika	Ankitsika	1	-	-
Total		**10**	**19**	**0**	**4**

In Ankarafantsika National Park, nine individuals were recorded, including eight adults made up of four normal pairs (n = 8 individuals) and one chick. In addition, two nests were known at the site (Table 5).

Tableau 5: *Results of a group survey on the presence of the Madagascar fish eagle in the Ankarafantsika site.*

Territory	**Adult individual**	**Nest**	**Chick**
Ravelobe	2	1	1
Doanibe	2	-	-
Tsimaloto	2	1	-
Komandria	2	-	-
Total	**8**	**2**	**1**

At Baie de Baly, five adult individuals were recorded, i.e. a normal pair with a nest and three solitary individuals. These five individuals are spread over four territories, one of which is on the edge of Lake Sarika and three in the mangrove swamps of the coastal zone (Table 6).

Tableau 6: *Results of a group survey on the presence of the Madagascar fish eagle in the Baie de Baly site.*

Territory	Type of ecosystem	Adult individual	Nest	Chick
Sari'ka	Lake	2	1	-
Kapoloza	Coastline	1	-	-
Vavanimoroleo	Coastline	1	-	-
Ankirakaraka	Coastline	1	-	-
Total		**5**	**1**	**0**

In total, in these four PAs, Nosy Hara, Sahamalaza - Radama Islands, Ankarafantsika and Baie de Baly, 34 territories, 69 individuals, including 63 adults divided into 29 normal pairs (n = 58 individuals) and five solitaries, and six chicks were recorded during the group surveys. Of the 29 pairs, 37.**9%** (n = 11) were reported to have nests (Table 7).

Tableau 7: *Summary of the results of the group survey in the four PAs of Madagascar National Parks.*

AP	**Territory**	**Adult individual**	**Couple**	**Couple with nest**	**Chick**
AMP Nosy Hara	16	31	15	8	1
APMC Sahamalaza	10	19	9	0	4
Ankarafantsika NP	4	8	4	2	1
Baly Bay NP	4	5	1	1	0
Total	**34**	**63**	**29**	**11**	**6**

III.2. POPULATION SIZE BY DIRECT OBSERVATION

Following direct observations in the field, 193 individuals were counted. They comprised 152 adults, seven sub-adults and 34 juveniles. These 152 adults were divided into 68 breeding pairs, 57 of which were normal (n = 114 individuals) and 11 polyandrous pairs (n = 33 individuals), i.e. a pair with two males, and five solitary individuals.

II.2.1. Population size in each type of ecosystem

The lake ecosystems and surrounding forests at the sites surveyed are home to 120 individuals, corresponding to 62.2% of the total population. Coastal areas are home to 62 individuals, representing 32.1% of the population counted. River ecosystems were home to only 11 individuals, or 5.7% of the total number of individuals counted at the 11 study sites (Figure 12).

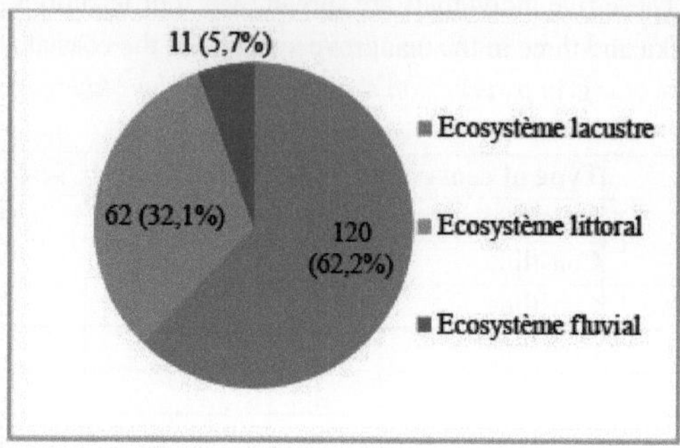

Figure 12: Population distribution of the Madagascar fish eagle in the three types of ecosystem studied.

111.2.1.1. Coastal ecosystem

All 62 individuals recorded in coastal areas were found in sites in the north-western part of the island, more precisely Nosy Hara, Sahamalaza - Radama Islands and Baie de Baly. These individuals are distributed as follows: 46 individuals from 23 pairs to two individuals, one solitary adult and 15 juveniles.

> ***AMP Nosy Hara and surrounding area***

At Nosy Hara, the visit was made with two AMP agents from 13 to 24 October 2016. A total of 20 individuals were counted, including 14 adults forming seven pairs of two individuals and six juveniles. Of these seven pairs, four were nesting on the marine islets and three on the coastal mangroves (Appendix 5). Apart from the individuals, one egg was found in a nest at Nosy Lakandava during our visit (Table 8).

Tableau 8: *Number of Madagascar fish eagles recorded in the Nosy Hara MPA and surrounding area.*

Localities	Territory	Latitude S	Longitude E	Nest	Couple	Individual of the couple	Juvenile
In the PA	Nosy Nosy Hara	12°15'10,2"	49°00'08,6"	2	1	2	1
	Nosy Lakandava	12°15'34,9"	48°57'22,8"	1	1	2	0
	Ankonkofaly	12°23'39,3"	48°45'42,8"	1	1	2	1
	Andovobe	12°24'25,4"	48°46'41,9"	1	1	2	1
	Fararano	12°24'59,8"	48°51'04,2"	1	1	2	1
Outside the PA	Nosy Valiha	12°23'14,4"	48°41'57,2"	1	1	2	1
	Nosy Mahanonoka	12°26'55,8"	48°39'52,6"	1	1	2	1
Total				**8**	**7**	**14**	**6**

In this table, we can see that 70% (n = 14) of the individuals inventoried in this site are located in the protected area and 30% (n = 6) are found on the periphery of the PA, especially on the marine islets.

> ***APMC Sahamalaza - Radama Islands and surrounding area***

The study was carried out in this PA and its surroundings between 28 October and 05 November 2016, i.e. over a period of nine days. In total, 33 individuals divided into 12 normal pairs (n = 24 individuals) and nine juveniles were inventoried. However, in 2017, the TPF team monitored this species at the same site. Their visit to the Ankitsika plot enabled them to discover two new normal pairs. As a result, the Sahamalaza - ìles Radama CMPA and its periphery are home to 14 normal pairs (n = 24 individuals) and nine juveniles (Table 9).

Tableau 9: *Madagascar fish eagle territories in Sahamalaza National Park - Radama Islands and surrounding area.*

Parcel	Territory	Latitude S	Longitude E	Nest	Couple	Individual of the couple	Juvenile
Ankitsika	Ankitsika	14°17'31,69"	47°59'30,60"	0	1	2	0
Ampasimbezo	Ankij animanjofo	14°16'39,1"	48°01'14,9"	1	1	2	1
Kapany	Anjiamalandy	14°05'11,2"	48°01'47,3"	1	1	2	1
	V avanankaramihely	14°06'47,3"	48°02'42,0"	1	1	2	1
	Ankerambalavo	14°04'12,5"	47°58'44,3"	0	1	2	1
	V avanambatomadoso	14°07'38,5"	48°02'23,1"	0	1	2	0
Sijoro	V avanambodidimaka	14°10'28,0"	47°56'45,7"	1	1	2	1
	Ambohikiriny	14°10'04,0"	47°56'53,4"	1	1	2	1
	Nosy Kalanoro	14°11'13,46"	48°1'35,85"	0	1	2	0
	Tranonkira	14°14'04,0"	47°56'08,9"	0	1	2	0
	Sijoro	14°07'50,6"	48°57'12,7"	1	1	2	0
Maromandia	Ankorobe	14°07'57,7"	47°59'49,3"	0	1	2	1
Outside the PA	Ambahofaho	13°57'04,6"	47°58'49,9"	1	1	2	1
	Andrafia	13°55'46,3"	47°57'08,18"	1	1	2	1
Total				**8**	**14**	**28**	**9**

Of the 14 pairs found, eight had nests and two were on the outskirts of the PA.

> ***Baly Bay National Park***

In this park, the survey was carried out by the TPF team and took place from 21 to 25 October 2016. In total, five (5) adult individuals forming two pairs with two individuals and a solitary were counted in the coastal zone. These two normal pairs were found respectively at Ankarananjy (16°01'31.4"S- 45°30'08.2"E) and Kapoloza .
(16°09'16,3"S - 45°03'47,9"E) while the solitary individual was located at Ambatomainty (16°04'40,8"S - 45°05'23,2"E). Apart from these individuals, one nest has been found in the coastal ecosystem of the Baie de Baly National Park.

III.2.1.2. Lake ecosystem

Of the 52 lakes visited, 30 were home to 120 individuals divided into 29 normal pairs (n = 58 individuals), 11 pairs with three individuals (n = 33 individuals), four solitary adults, i.e. 95 adult individuals, 18 juveniles and seven sub-adults (Figure 13).

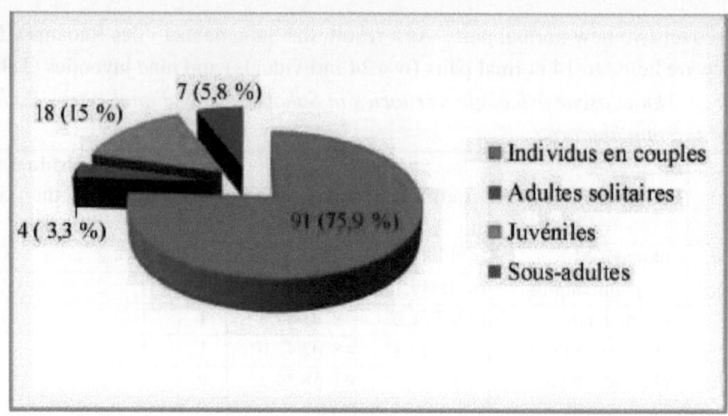

Figure 13: Population size of *Haliaeetus vociferoides* in lake ecosystems.

> **Lake Tseny and Ankarafantsika National Park**

In Lake Tseny, a solitary adult and a nest were found between the geographical coordinates 15°40'41.8"S - 47°50'33.1"E.

The census in Ankarafantsika National Park was carried out jointly by the TPF team and the National Park team from 15 to 19 October 2016. A total of nine (9) individuals were counted, including eight adults forming four normal pairs and one juvenile. A pair with a nesting juvenile was found on the edge of Lake Ravelobe (16°18'7.3 "S - 46°49'11.0 "E). The other three pairs, the first of which had a nest, were located respectively in lakes Komandria (16°34'72.2 "S- 46°65'64.2 "E), Doanibe (16°34'28.9'S' - 46°73'84.7"E) and Tsimaloto (16°22'33.7"S - 47°13'74.7"E).

> **Baly Bay National Park**

In the Baie de Baly National Park, the lake ecosystems are home to six (6) individuals forming three normal pairs. Of these three pairs, the first was located in Lake Sarka between the geographical coordinates 15°58'42.4"S - 45°11'59.8"E and the last two are located in Ankoro between the geographical coordinates 16°09'21.3"S - 45°10'58.9"E and 16°09'43.1"S - 45°10'30.1"E.

> **Besalampy district**

The Besalampy wetlands were surveyed by three technicians from Mandrozo PA between 12 and 16 July 2016. Six (6) individuals including two normal breeding pairs, a stray adult and a chick were found. Four individuals, i.e. a normal pair, a nesting chick and a solitary adult were found in the Amparihy wetland (16°41'36.1"S - 44°49'38.0"E). The second normal pair was located in the Sahapy wetland (16°31'50.7"S - 44°42'31.7"E).

> **Maintirano District**

The Tambohorano wetland was visited between 20 and 25 July 2016. This enabled us to find 15 individuals including 11 adults, three juveniles and one sub-adult in Lake Mandrozo and the surrounding satellite lakes. These 11 adult individuals were divided into five normal pairs (n = 10 individuals) and one solitary individual. Of these five pairs, one did not have a nest. Two territories, including Mahiarere and Nosy Sarotra, lie within the Mandrozo PA (Table 10).

Tableau 10: *Results of the Madagascar fish eagle count in the Tambohorano wetland,*

Maintirano District.

Location	Territory	Latitude	Longitude	Nest	Cpl	Ind cpl	Adlt slt	Juv	Ss- adlt
In Mandrozo PA	Mahiarere	17°34'11,5"	44°08'31,4"	1	1	2	-	-	-
	Nosy Sarotra	17°33'05,5"	44°04'56,0"	-	-	-	1	-	-
In the satellite lakes	Bezozoro	17°07'17,3"	44°14'39,8"	-	1	2	-	-	1
	Ambalarano	17°08'05,6"	44°14'41,8"	1	1	2	-	1	-
	Andranovorinakoay	17°12'57,0"	44°16'00,9"	1	1	2	-	1	-
	Antevamena	17°36'56,2"	44°10'14,1"	1	1	2	-	1	-
	Tota			**4**	**5**	**10**	**1**	**3**	**1**

Cpl: *Couples,* ***Ind cpl****: Individuals of the couple;* ***Adlt slt****: Solitary adults;* ***Juv****: Juveniles;* ***Ss-adlt****: Sub-adults*

Antsalova district

The research in the Antsalova district took place between 30 July and 20 August 2016, coinciding with the period of the second nest observation visit of the year. In total, 14 of the 20 lakes visited are home to the Madagascar Fish Eagle. In these 14 lakes, 64 individuals were counted, including 50 adults, eight juveniles and six sub-adults. The 50 adults were made up of 19 pairs, including 11 normal pairs (n = 22 individuals), nine polyanders with two males (n = 27 individuals) and one solitary adult.

The Tsimembo Manambolomaty Complex PA is home to 24.4% of the entire population found in the 11 study sites, i.e. 47 individuals including 36 adults, six juveniles and five sub-adults. Of these 47 individuals, 36 corresponding to 49.3% of the Fish Eagle population in the Antsalova district (n = 73 individuals), were located in the complex of three Manambolomaty lakes (Ankerika, Soamalipo and Befotaka). These individuals can be broken down into four pairs with two individuals (n = 8 individuals), six pairs with three individuals (n = 18 individuals), five sub-adults and five juveniles. Apart from these three lakes, 11 individuals forming two normal pairs (n = 4 individuals), two polyanders (n = 6 individuals) and one juvenile were found in the satellite lakes located within the PA Tsimembo Manambolomaty Complex. In this PA, all the nests of this species were located except that of the pair at Beora (Table 11).

Tableau 11: *Number of Madagascar fish eagles inventoried in the PA Tsimembo Manambolomaty district of Antsalova.*

Territories	Latitude	Longitude	Nest	Cpl	Ind cpl		Juv	Ss-adlt
Nosindambo	19°01'03,1"	44°24'31,1"	1	1	2		-	-
Befotaka 2	19°01'36,9"	44°23'58,9"	1	1	3		1	-
Befotaka 3	19°00'26,4"	44°24'32,1"	1	1	2		-	-
Soamalipo 1	19°01'17,5"	44°25'09,9"	1	1	3		1	-
Soamalipo 2	18°59'09,8"	44°25'18,5"	1	1	3		-	1
Soamalipo 3	19°01'29,9"	44°25'58,8"	1	1	2		1	-
Ankerika 2	19°02'09,7"	44°26'54,8"	1	1	3		1	1
Ankerika 3	19°00'28,0"	44°27'44,5"	1	1	3		-	1
Ankerika 4	19°00'34,1"	44°27'09,9"	1	1	3		1	1
Ankerika 5	19°04'06,5"	44°27'31,3"	1	1	2		-	1
And aranolava-nord	19°00'25,9"	44°20'56,4"	1	1	3		1	-
Antsamaky	19°02'52,3"	44°21'43,1"	1	1	2		-	-
Beora	18°52'43,8"	44°22'03.4"	-	1	2		-	-

Masama	18°51'01,9"	44°27'16,4"	1	1	3		-	-
Total			**13**	**14**	**36**		**6**	**5**

***Cpl:** Couples, **Ind cpl**: Individuals of the couple; **Ju**: Juveniles; **Ss-adlt**: Sub-adults*

Apart from the Tsimembo Manambolomaty Complex PA, seven of the 11 lakes visited are home to the Madagascar Fish Eagle. In these seven lakes, 17 individuals were counted, including 14 adults, two juveniles and one sub-adult. These 14 adults were divided into six breeding pairs, five of which were normal (n = 10 individuals), one polyandrous with two males (n = 3 individuals) and one solitary individual. Of these six pairs, four were recorded with nests (Table 12).

Tableau 12: *Numbers of Madagascar Fish Eagle counted in satellite lakes outside the Tsimembo Manambolomaty Complex PA, Antsalova district.*

Lakes	Latitude	Longitude	Nest	Cpl	Ind Cpl	Adlt sol	Juv	Ss-adlt
Ambohibary	18°48'17,9"	44°24'41,5"	-	-	-	1	-	-
Ampozabe	19°08'41,9"	44°39'49,5"	-	1	2	-	-	-
Andranolava-sud	19°05'27,5"	44°24'43,1"	-	1	2	-	-	-
Antsalaza	18°52'49,1"	44°34'51,2"	1	1	2	-	-	-
Antsahafa	18°48'42,2"	44°28'38,1"	1	1	2	-	1	-
Antsakotsako	19°06'10,5"	44°32'42.4"	1	1	2	-	-	1
Antsohale	19°07'41,4"	44°35'05,3"	1	1	3	-	1	-
Total			**4**	**6**	**13**		**2**	**1**

***Cpl:** Couples, **Ind cpl**: Individuals of the couple; **Ju**: Juveniles; **Ss-adlt**: Sub-adults*

> **Belo-sur-Tsiribihina district**

In this site, 13 lakes were visited between 13 and 22 July 2016. During these 10 days of observations, 10 individuals forming three normal pairs (n = 6 individuals), one polyandrous pair and one juvenile were inventoried. These three normal pairs were found in three lakes, including Andranovoribe (19°12'11,6"S - 44°50'28,2"E), Ankevo (19°12'30,6"S - 44°31'52,3") and Kimanomby (19°46'53,7"S - 44°43'51,7" E). The polyandrous pair and a juvenile are located in Lake Bejijo (19°13'23.5"S - 44°32'26.0"E). Apart from these individuals, two nests were spotted at Andranovoribe and Bejijo.

> **Miandrivazo District**

In Miandrivazo District, the survey took place from 13 to 22 December 2016, i.e. over a period of 10 days. During this period, nine lakes were visited. Nine (9) individuals, including six adults and three juveniles, were inventoried. These nine individuals were distributed between three lakes: Andranomazava (19°43'26,1"S - 45°18'52,6"E), Asonjo (19°50'36,5"S - 45°26'30,9"E) and Badika (20°09'14,3"S - 45°31'51,9"E), forming three normal breeding pairs with one juvenile each. The nest of the Badika pair was not found because the nesting territory had been burnt about a month before our visit.

111.2.1.3. River ecosystem

Eleven (11) individuals divided into three normal pairs (n = 6 individuals), one polyandrous pair (n = 3 individuals) and two juveniles were discovered in this type of ecosystem. Of these three normal pairs, two were located in the Antsalova district, the first at Androibe (18°42'02,2 "S - 44°31'49,1 "E), on the banks of the Soahany river, while the second with a

juvenile was spotted on the banks of the Manambolo river, near Bekopaka (19°08'23,3 "S - 44°48'38,2 "E). The third pair of two individuals was found along the Tsiribihina river, more precisely at Ambatomainty (19°44'25,0" S- 44°48'38"E).
44°55'20,0"E). Finally, a polyandrous pair with a juvenile is also located along the Soahany river, between the geographical coordinates 18°47'30.0"S - 44°19'10.0"E.

III.2.2. Total population numbers in each study site

Based on the above results, it was possible to determine the total size of the Madagascar fish eagle population for each study site (Table 13).

Tableau 13: *Population size of the Madagascar fish eagle at each study site.*

Website	Ecosystem			Total	Rates.
	Lake	Coastline	River		
Nosy Hara	-	20	-	20	0,104
Sahamalaza - Radama Islands	-	37	-	37	0,192
Baly Bay	6	5	-	11	0,057
Lake Tseny	1	-	-	1	0,005
Ankarafantsika	9	-	-	9	0,047
Besalampy district	6	-	-	6	0,031
Maintirano District	15	-	-	15	0,078
Antsalova district	64	-	9	73	0,378
Belo-sur-Tsiribihina district	10	-	2	12	0,062
Miandrivazo District	9	-	-	9	0,047
Mangoky-Ihotry complex	-	-	-	0	0
Total	**120**	**62**	**11**	**193**	**1,000**

***Propp**: Proportion of the site in relation to the total observed.*

It can be seen from Table 12 that 10 of the 11 sites surveyed are home to the Fish Eagle. The Antsalova District is considered to be the most abundant site for the Fish Eagle population in Madagascar, where the population represents 37.8% (n = 73 individuals) of the total population (Figure 14). On the other hand, the CIM PA, which includes the Manja-Morombe wetlands, is not home to any individuals of Madagascar fish eagle.

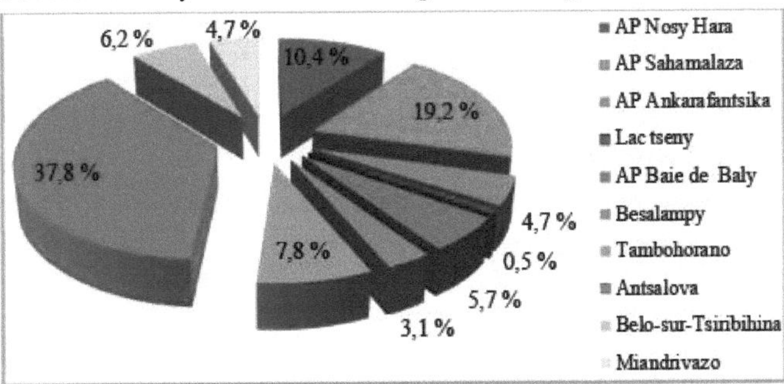

Figure 14: Population size of *Haliaeetus vociferoides* by site.

111.2.3. Comparison between group survey results and results of direct observations.

Group surveys were carried out only in the four protected areas of Madagascar National Parks, such as Nosy Hara MPA, Sahamalaza - iles Radama CMPA, Baie de Baly NP and Ankarafantsika NP. In general, the results of direct observations are higher than those of surveys. However, the results vary from site to site. (Table 14).

Tableau 14: *Comparative summary of the results of surveys and direct observations at the level of the four PA of Madagascar National Parks.*

Sites	Results of group surveys				Results of direct observations			
	Territory	Individual	Couple	Couple with nest	Territory	Individual	Couple	Couple with nest
AMP Nosy Hara	16	32	15	8	7	20	7	7
APMC Sahamalaza	10	23	9	0	14	37	14	8
Ankarafantsika NP	4	9	4	2	4	9	4	2
Baly Bay NP	4	6	1	1	6	11	5	2
Total	34	69	29	11	31	77	30	19

In the Nosy Hara MPA and its periphery, the numbers recorded during the group survey were higher than those obtained during direct observations. Only 43.7% (n = 7) of the number of territories, 62.5% (n = 20) of the number of individuals and 46.7% (n = 7) of the number of pairs recorded in the survey were observed in the field. On the other hand, in the Sahamalaza - Radama Islands and surrounding area CMPA, 140% of the total number of pairs were observed in the field.

(n = 14) of the number of territories; 160.9% (n = 37) of the individuals and 155.5% (n = 14) of the pairs reported during the survey were counted by direct observation. The same is true for direct observation at Baie de Baly NP, where the proportions are 150% (n = 6 territories), 183.3% (n = 11 individuals) and 500% (n = 5 pairs) respectively. At Ankarafantsika NP, the two results coincide perfectly.

As for the results relating to reproduction, i.e. the number of pairs with nests and the number of juveniles, the survey yielded only very small numbers compared with direct observations. As a result, at all four sites, 172.7% (n = 19 pairs) of the total number of pairs with nests recorded during the survey and 266.7% (n = 16) of the number of juveniles counted during the survey were inventoried in the field. The highest number of pairs with nests was found in Sahamalaza - Radama Islands, as none of the 8 pairs with nests observed in the field were recorded during the survey.

111.3. DISTRIBUTION OF WORKFORCE BETWEEN PROTECTED AND UNPROTECTED AREAS

Generally speaking, protected areas contain more *Haliaeetus vociferoides* than unprotected areas. In these protected areas, 123 individuals were counted, corresponding to 63.7% of the total number of individuals counted, compared with 70 individuals in unprotected areas, representing 36.3% of the total population counted.

At the Baie de Baly and Ankarafantsika sites, all the individuals recorded were found in protected areas. However, at the Besalampy and Miandrivazo sites, no individuals were found within the PAs. At Nosy Hara and Sahamalaza, 70% (n = 14) and 83.8% (n = 31) of the total number of individuals were found in PAs. In the Antsalova district 64.4% (n = 47) of the individuals surveyed live in the Tsimembo Manabolomaty PA (Figure 15).

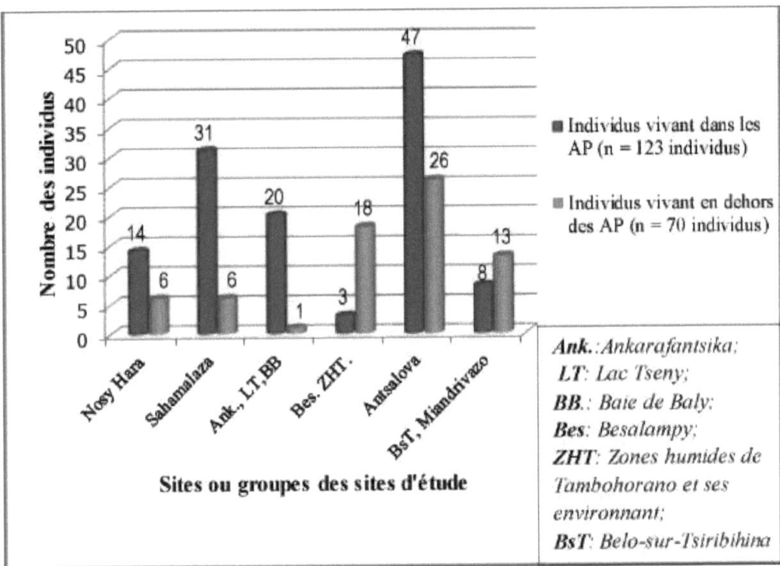

Figure 15: Number of individuals recorded living inside and outside PAs, by study site.

In addition, statistical analysis (chi-square test: $\chi^2_{cal} = 37{,}57$; $ddl = 5$; $\propto = 0{,}05$; $\chi^2_{tab} = 9{,}49$ et $p < 0{,}0001$) showed that the distribution of numbers in the Madagascar fish eagle population between these protected and unprotected areas is heterogeneous.

111.4. GEOGRAPHICAL DISTRIBUTION OF *Haliaeetus vociferoides*

The area of occurrence of the Madagascar Fish Eagle extends across the vast plains along the west coast of the Grande Ile, from Nosy Hara in Antsiranana, in the north, to the District of Belo-sur-Tsiribihina, in the south. Inland, the distribution of this species extends as far as the districts of Maevatanana (pers. com. AMPA researcher, 2016) and Miandrivazo. Nevertheless, the main habitats are found in the wetlands of Antsalova, Tambohorano, the bays of Mahajamba and Moramba, in the PAs Baie de Baly, Ankarafantsika, Sahamalaza and Nosy Hara, and in the Mitsio Archipelago.

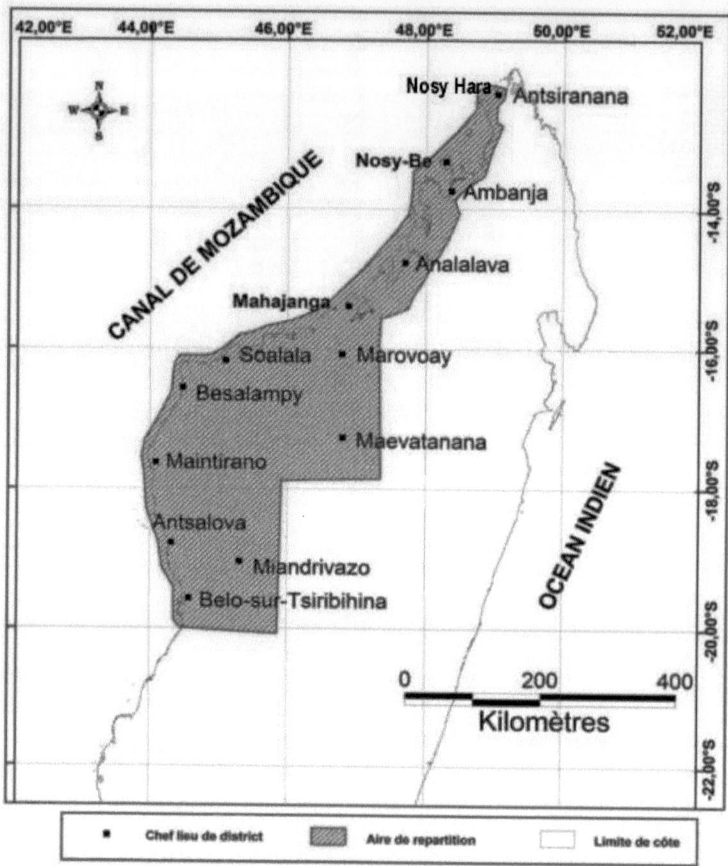

Figure 16: Updated map of the geographical distribution of *Haliaeetus vociferoides* in France.
2016 (Rasolonjatovo, 2018)

111.5.CHANGE IN POPULATION BETWEEN 2006 AND 2016

Of the 11 sites in the present study, only Lake Tseny was not one of the target sites for the 2005 - 2006 census. At that time, the same sites visited during the present study were home to 168 individuals of the Madagascar fish eagle population (Extract from Razafimanjato, 2011). At three sites, Antsalova, Besalampy and especially Belo-sur-Tsiribihina, the rates of population change are negative. We can therefore deduce that populations at these sites have been declining between 2006 and 2016. However, the populations of the other sites are increasing. The highest rate of increase was observed at the Ankarafantsika site, where it was equal to 2 and the population grew by an average of 11.6% individuals per year (Table 15).

***Table 15:** Population trend of the Madagascar fish eagle in the study sites.*

Sites	*EQ*	*Er*	$E_r - E_o$	***T***	*Ea*	***Ra (%)***	***Sense of evolution***
Nosy Hara	*10*	*20*	*10*	*1*	*0,072*	*7,2*	*Growing trend*

Sahamalaza-iles-Radama	*26*	*37*	*11*	*0,4*	*0,036*	*3,6*	*Growing trend*
Baly Bay	*5*	*11*	*6*	*1,2*	*0,082*	*8,2*	*Growing trend*
A nkarafantsika	*3*	*9*	*6*	*2*	*0,116*	*11,6*	*Growing trend*
Besalampy	*9*	*6*	*-3*	*-0,33*	*-*	*-*	*Declining trend*
Maintirano	*12*	*15*	*3*	*0,25*	*0,022*	*2,2*	*Growing trend*
Antsalova	*77*	*73*	*-4*	*-0,05*	*-*	*-*	*Declining trend*
Belo-sur-Tsiribihina	*19*	*12*	*- 7*	*-0,37*	*-*	*-*	*Declining trend*
Miandrivazo	*7*	*9*	*2*	*0,28*	*0,025*	*2,5*	*Growing trend*
All sites	***168***	***192***	***24***	***0,14***	***0,013***	***1,2***	***Growing trend***

E_o : Population size in 2006. E_r : Population in 2016. ***t****: Rate of change in population size. R_a : Average annual growth rate.*

Across all the study sites, the population grew by an average of 1.3% per year. However, the variation in population size between 2006 and 2016 was not statistically significant (Wilcoxon test: $N = 18$; $p = 0.31 > 0.05$). This means that, at all these sites, the *Haliaeetus vociferoides* population has not shown an expressive increase over the last decade.

111.6. EAGLE POPULATIONS IN ANTSALOVA AND IN THE TAMBOHORANO WETLAND

111.6.1. Population change

As a result of annual monitoring of Madagascar fish eagle populations at the Antsalova site (site 1) and the Tambohorano wetland and surrounding area (site 2), we have obtained data enabling us to describe its variation over the last decade (Appendix 7).

At site 1, the most abundant period for Madagascar fish eagles was 2010-2011, when the population reached 87 individuals. The population abundance at this site was lowest in 2015 (n = 55 individuals). At site 2, the years with the lowest numbers of Madagascar fish eagles (n = 10 individuals) were 2007 and 2013. However, 2009 was the year with the highest abundance of Madagascar Fish Eagle, both in site 2 and in the two sites as a whole. In that year, the population numbers were 20 and 103 individuals respectively. The year 2015 corresponds to the lowest total population size for these two sites, i.e. 60 individuals (see Appendix 6).

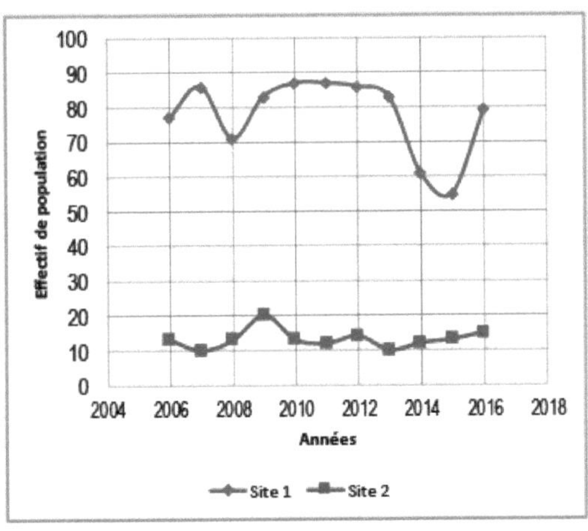

Figure 17: Variation curves for Madagascar fish eagle populations in the two TPF intervention sites, as a function of time between 2006 and 2016.

17.6.2. . Reproductive success

At site 1, the breeding success rate averaged 75.8% small per brood (n = 214 active nests, sd = 10.41). During the period 2006 - 2016, a total of 300 eggs were recorded in these 214 active nests. Of these, 62% (n = 186 eaglets) hatched and 162 or 87.1% of the chicks reached the age of first fledging (Appendix 6).

At site 2, this rate varies on average around 66.6% (n = 35 active nests; sd = 25.3). In fact, 36 chicks hatched from the 67 eggs laid. Of these 36 chicks, 21 fledged successfully (Appendix 6).

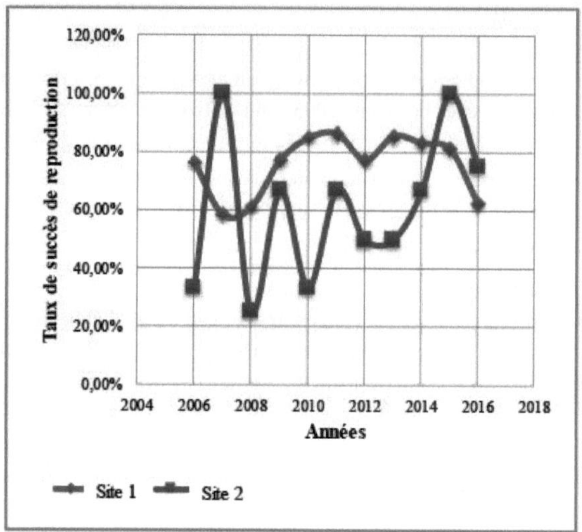

Figure 18: Variation curve for breeding success at the two TPF intervention sites between 2006 and 2016.

In site 1, the year with the lowest breeding success rate was 2007 (58.3% chicks per brood), while the highest rate was in 2011 (86.4% chicks per brood). In the Tambohorano wetlands, the lowest rate was in 2008 (25% chicks per brood) but the highest was in 2007 and 2015, when the breeding success reached 100% chicks per brood.

18.6.3. . Comparison of breeding success between the two sites

> ***Comparison of egg laying between site let site 2***

There was a significant difference between the number of eggs laid in site 1 and site 2 (Mann-Whitney test: U = 223; N = 22; $p < 0.0001 < 0.05$). The average number of eggs laid in the Antsalova district was higher than in the Tambohorano wetland, Maintirano district. These averages are 27.27 (n = 300; sd = 4.80) and 6.09 (n = 67; sd = 0.94) eggs per year respectively (Figure 19).

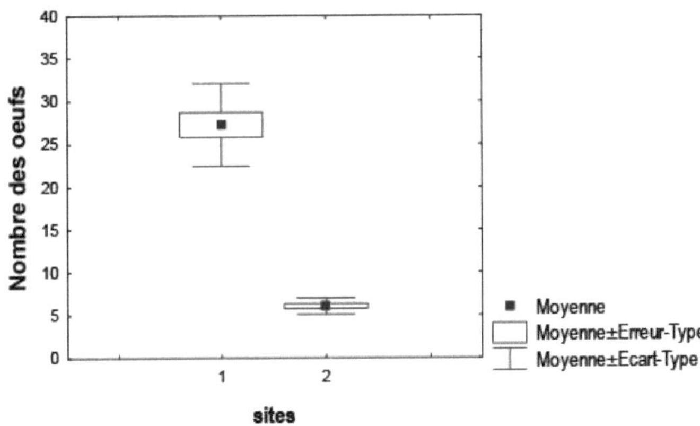

Figure 19: Average number of eggs laid at the two sites between 2006 and 2016.

> ***Comparison of hatching between site 1 and site 2***

The difference in the mean number of eggs hatched between site 1 and site 2 was significant (Mann-Whitney test: U = 226.024; N = 22; $p < 0.0001 < 0.05$). Site 1 produced significantly more chicks than site 2 (Figure 20).

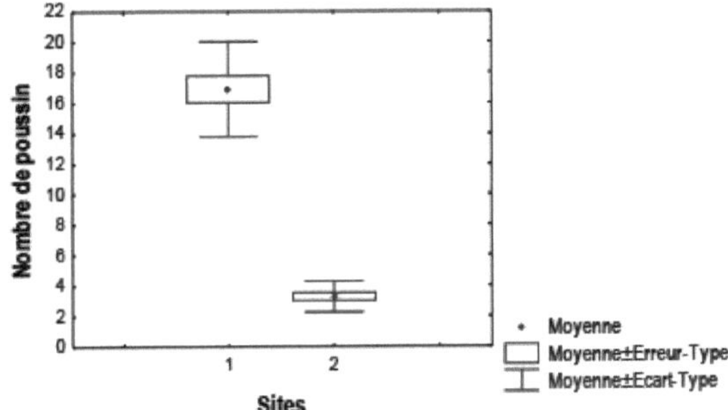

Figure 20: Chick averages at the two sites between 2006 and 2016.

> ***Comparison of shipments between site 1 and site 2***

At site 1, the average number of chicks fledged during the decade was 14.72 (n = 162; sd = 3.29). In site 2, on the other hand, it was 1.91 (n = 21; sd = 0.83). The difference between these two means is significant (Mann-Whitney test: U = 227.333 ' TU = 227,333; N = 22; p < 0,001 < 0,05), comme), as shown in Figure 21.

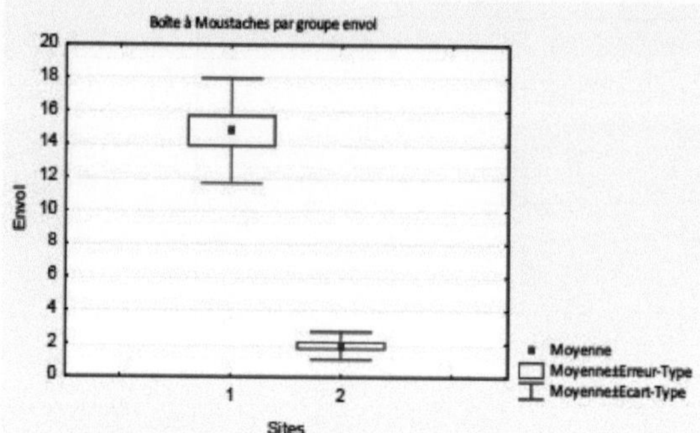

Figure 21: Average number of chicks fledged at the two sites between 2006 and 2016.

In short, each stage in the reproduction of the Madagascar fish eagle (egg-laying, hatching and fledging) has differed between the two sites over the last decade.

III.7. NIDIFICATION

111.7.1. Number of nests

Following systematic observations, 47 of the 54 nests counted were found to be intact. Of these 47 nests, 29, 15 and 3 were located in the continental, coastal and riverine ecosystems respectively. At the three sites of Lac Tseny, Baie de Baly and Besalampy, only one nest was found at each site. The nest at Lac Tseny had been abandoned, while at the other two sites the nests were all old but intact.

Table 16: *Some parameters concerning Madagascar fish eagle nests in the 2016 biological year.*

Study sites	Nests found	Nests abandoned	Intact nests		Average distance between s two neighbouring nests (km)
			New	Ander	
Nosy Hara	8	1	1	6	6 (n = 7; sd = 2.87)
Sahamalaza	10	2	3	5	5.24 (n = 8; sd = 5.08)
Tseny and Ankarafantsika	3	1	-	2	18 (n = 2)
Baly and Besalampy bays	2	-	-	2	-
Maintirano	4	-	2	2	11.59 (n = 4;sd = 3.26)
Antsalova (3 lakes)	10	-	1	9	1.84 (n = 10; sd = 0.96)
Antsalova (satellite lakes)	13	3	3	7	9.17 (n = 10; sd = 4.61)
Bello-sur-Tsiribihina	2	-	-	2	4,825 (n = 2)
Miandrivazo	2	-	-	2	18,8 (n = 2)
Total	**54**	**7**	**10**	**37**	-

Of the 47 nests in an intact state, 78.7% (n = 37) are in an old state. The smallest average distance between two neighbouring nests (1.84 km; with n = 10 and sd = 0.96) is found in the complex of three lakes in the Tsimembo Manambolomaty PA in the Antsalova district. The most distant nests are found in Miandrivazo and are 18.8 km apart.

111.7.2. Change of nest location

During this study, nine (9) cases of nest change were found. The factors responsible for this change were mainly anthropogenic, such as the frequency of visitors or fishermen passing through the nesting site, the felling of nest-bearing trees and the clearing of the nesting area. Nest collapse and reproductive failure are also factors in nest change for this species. The distances separating new and old nests found did not exceed one kilometre and varied between 70.4 and 995 m, i.e. an average of 294.71 m (n = 9; sd = 278.50) (Table 17).

Table 17: *Distances between new and old nests and factors influencing nest change.*

Nest change territories	Distance between new and old nests (m)	Factors in nest change
Nosy Hara	209	Visitor frequency
Ambavanambodidimaka	170	Frequency of passage of fishermen
Andrafia	400	Frequency of passage of fishermen
Ankij animanj ofo	127	Nest tree cutting
Ambalarano	180	Nest tree cutting
Somalipo 3	233	Nest fall
Andranolava-nord	995	Reproductive failure
Antsamaky	268	Nest fall
Bekopaka	70,4	Clearing land for agriculture
Mean ± standard deviation	**294.71 (sd = 278.50)**	

111.7.3. Nest location and characteristics

In general, the nest of the Madagascar fish eagle is located on the top of trees that emerge very close to a permanent source of water. Supported by 3, 4 or even 5 branches forming a fork, the nest is shaped like a more or less ovoid cup with the opening pointing skywards (Figures 23 and 24). The nest is built using small branches ranging from 1.5 to 3 cm in diameter and 30 to 65 cm in length, as well as green leaves. The complete structure measures an average of 124.25 cm (n = 4; sd = 4.65), with the log measuring 70.75 x 53 cm (n = 4; sd = 3.95 x 5.72). The average height and depth are 55.5 cm (n = 4; sd = 5.44) and 13.75 cm (n = 4; sd = 8.53). In all ecosystems (coastal and continental), the height of the nests in relation to the ground varied from 5 to 24 m, i.e. an average of 13.91 m (n = 40; sd = 4.3). However, in the continental zone, nests were built between 12 m and 24 m high, with an average height of 16.42 m (n = 26; sd = 3), compared with 9.25 m (n = 14; sd = 2.41) in the coastal zone, where nest heights varied between 5 m and 13 m. According to statistical analysis (t-test: **t** = 2.042; **sd** = 30; **p** < 0.0001 < 0.05), the difference between these two nest height averages is significant, as shown in Figure 22.

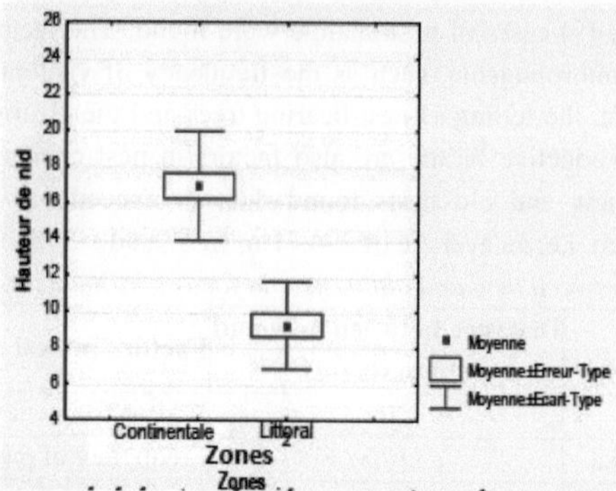

Figure 22: Average nest height in relation to the ground.

Figure 23: Madagascar Fish Eagle nest in the continental zone, at Ankerika 3 Rasolonjatovo, 2018).

111.7.4. Nesting tree characteristics

With the exception of two nests on the rocky islets Nosy Hara and Lakandava, which are built on *Securinega perrierii* (Hazomena*)* and *Terminalia mantaly* (Mantaly Beravina*),* all the intact nests found in coastal areas (n = 16) are built on *Sonneratia alba* (Farafaka*)* belonging to the Lythraceae family. This species is known locally as "Songere" in the south-west region of Madagascar (Figure 24). However, in the continental environment, the 28 nests studied were built on 11 different tree species, dominated by *Tamarindus indica* (Kily) followed by *Colvillea racemosa* (Sarongaza), all belonging to the Fabaceae family (Table 18).

Table 18: *Madagascar fish eagle nest tree species from the 2016 biological year.*

Zones	Families	Species	Common names	N.e
Coastal	COMBRETACEAE	*Terminalia mantaly*	Mantaly	1
	LYTHRACEAE	*Sonneratia alba*	Farafaka, Songere	14
	PHYLLANTHACEAE	*Khaya madagascariensis*	Hazomena	1
Continental	ANACARDIACEAE	*Mangifera indica*	Manga	1
	BIGNONIACEAE	*Streospermum bernieri*	Vavaloza	1
	BOMBACACEAE	*Adansonia sp*	Reniala	4
	DIDIEREACEAE	*Terminalia trricristata*	Taly	2
	FABACEAE	*Tamarindus indica*	Kily	8

		Colvillea racemosa	Sarongaza	5
		Cordyla madagascariensis	Anakakaraka	1
	RUBIACEAE	*Breonadia salicina*	Sohihy, Soaravina,	3
	MORACEAE	*Broussonetia greveana*	Vory, Somely, Moramana	1
		Ficus coculifolai	Adabo	1
	Unidentified	Unidentified	Unidentified	1
Total	**10 families identified**	**14 species**		**43**

N.e: *Number of samples.*

Figure 24: Madagascar fish eagle nest built on the Farafaka in the coastal zone at Nosy Valiha (Rasolonjatovo, 2018).

As for the diameter at breast height (DBH) of the nesting trees, measurements show that it varies between 26.1 and 186 cm with an average of 66.06 cm (n = 40; sd = 36.37). However, a difference in Dhp was observed between the continental environment and the littoral zone. In the continental zone, Dhp ranged from 32.2 to 186 cm, with an average of 79.31 cm (n = 26; sd = 38.87), whereas in the coastal zone Dhp values ranged from 26.1 to 53.4 cm, with an average of 41.48 cm (n = 14; sd = 7.76). This difference in mean nest tree Dhp between these two environments is statistically significant (Mann-Whitney test: U = 1019.545; N = 62; $p < 0.0001 < 0.05$).

111.7.5. Characteristics of habitats surrounding nests

111.7.5.1. Description of nesting site

In general, the Madagascar fish eagle nests in a more or less dense forest, near a permanent body of water rich in fish or on rocky or forested islands and in a more or less intact mangrove swamp. Sometimes nests can be found built on the remains of large trees in degraded areas. Details of nesting territories are shown in Appendix 5.

111.7.5.2. Nest distance from permanent water body

In the continental environment, the distance between the nest and the permanent water body varied from 0 (zero) to 802 m, i.e. an average of 176 m (n = 26; sd = 206). Of these 26 nests, 57.7% (n = 15) were less than 100 m from the permanent water body. The furthest nest from the water, 802 m from the lake, was at Antsamaky in the Tsimembo Manambolomaty PA, Antsalova district.

In the complex of three lakes in the Tsimembo PA, the distance between the nest and the water varies from 18 to 292 m (Appendix 3).

111.7.5.3. Nest distance from disturbing elements

Twenty-two (21) of the thirty-six (36) nests studied were located less than 2 km from fishermen's or farmers' camps and villages. The pair at Amparihy appears to be the most disturbed since the tree carrying the nest is located 7 m from the camp (Figure 25). The results concerning the distances separating the nest tree from the disturbing elements are shown in Appendix 5.

For the purposes of this study, no nests of other raptor species were found near Madagascar fish eagle nests.

Figure 25: A camp at the nesting site in Amparihy (Rasolonjatovo, 2018).

III.8. THREATS AND PRESSURES

111.8.1. Threats and pressures on the species and its forest habitat

Following surveys and direct observations in the field, we found that the hunting of juveniles and the belief held by local people that this species was a bad omen were having a negative impact on the Madagascar fish eagle population. Among the anthropogenic actions, we noted the following:

- **bush fires:** these are either provoked or produced accidentally. In the first case, fires are used to renew pastures and collect honey. In the second case, the fires are caused by the cleaning of plots of land before sür-brulis cultivation (Figure 26).

Figure 26: Trail of fire in the nesting area at Andranovorinankoay (Rasolonjatovo, 2018).

- **Felling large trees:** dugout canoes require large-diameter trees of well-defined species that are light and easy to work, such as *Givotia stipuliflora* (Sefo), *Gyrocarpus americanis* (Mafay), *Commiphora guillaumii* (Arofy) and others. This manufacture has been noticed in all forests, even in protected areas as well as in nesting territories such as the case of Ankerika 2 at 45 m and Soamalipo 3 at 27 m of nest tree. Large trees such as *Broussonetia greveana* (Vory), *C. guillaumii* and *Dalbergia* sp (Manary) are also used to make planks for furniture.

Figure 27: Felled rosewood (*Dalbergia* sp) encountered 23 m from nesting tree at Ambalarano (Rasolonjatovo, 2018).

- **Roving livestock:** this type of pressure was encountered in all the forest habitats surveyed. Trampling by cattle in the forests around the lakes inhibits natural regeneration. The soil is no longer completely covered and its particles are easily eroded, leading to the risk of silting up the lakes.

In addition, the development around the nesting area and the cutting down of large trees on the lakeshore, as well as the conversion of the lakeshore into rice fields, have also been observed, especially in unprotected areas such as Amparihy, Sahapy, Andranovorinankoay and Ambalarano. All these factors lead to a loss of habitat for this species.

111.8.2. Threats and pressures on lakes

Lakes are favourable wetlands for the survival of many species of waterfowl. This is due not only to their nutritional resources but also to the grassy vegetation on the banks, which provides suitable nesting places for the species. From the observations made in the field, we have noted that the specific social, economic and environmental problems of the lakes included in the protected areas are fairly or less serious compared with those in unprotected areas.

In protected areas such as Tsimembo Manambolomaty, Mandrozo, Baie de Baly and Ankarafantsika, disturbance remains low and ecological integration is still maintained. The contents of the specifications are generally respected by local managers. Most fishermen have complied with the regulations on fishing, the rite such as "loa-drano" or "teabony" and the related taboos. However, some migrant fishermen do not respect either the regulations or the customs, hence the practice of illegal fishing during the closed season, the existence of night fishing even in the hard cores, and the presence of an illegal camp on the edge of the lake.

On the other hand, in unprotected areas, the lakes are threatened by overfishing, i.e. fishing is practised even during the resource renewal period in the case of Amparihy, Sahapy, Bezozora, Andranomazava and Badika. In addition, there are too many fishermen in lakes Amparihy, Sahapy and Badika. In addition, the use of nets over 250 m long has been noted in lakes

Amparihy, Sahapy and Andranomazava. The customs and cultures associated with the lakes are no longer respected, for example in Amparihy and Sahapy where the taboo prohibits the transport of hens and salt across the lakes, but the migrants no longer respect them. In addition, nets with mesh that is too tight (like mosquito netting) have been found for fishing *Stolephorus heterolobus* (Varilava), except at Andranomazava where the water is very deep. In addition, at Amparihy, Sahapy, Ambalarano and Badika, the lake shore has been transformed into rice fields and pastureland. Moreover, we found the spread of the invasive plant species *Eichhornia crassipes* (Tsikafonkafona) especially in Lake Asonjo and Andranomazava, and the formation of *Phragmites sp* covering a large part of lakes Amparihy, Sahapy, Bezozoro and especially Badika. This
poses a threat to the lakes and hinders the fish eagle's prey drive.

Figure 28: *Stolephorus heterolobus* fishing nets in Asonjo (Rasolonjatovo, 2018).

Figure 29: Invasion of *Eichhornia crassipes* at Lake Asonjo (Rasolonjatovo, 2018).

111.8.3. Threats and pressures in the coastal zone

In general, the human activities and economic pressures observed in the mangrove ecosystem can be summarised by operating systems with divergent rationales:

- **mangrove harvesting for energy wood and building materials**: mangrove trees are used by the entire coastal population for building and, to a lesser extent, for agriculture.
for heating (Figure 30).

Figure 30: Heap of mangrove wood cut for charcoal production at Ambahofaho, periphery of Sahamalaza National Park (Rasolonjatovo, 2018).

- **transformation into cultivated fields:** this case was encountered mainly in the Tsiribihina estuary mangrove around Soarano.
- **salt extraction:** this case was found in Besalampy and Belo-sur-mer:
- **industrial shrimp fishing:** in the study area, this industry can be found in Baie de Baly, Besalampy, Belo-sur-Tsiribihina and Belo-sur-mer. In general, aquaculture has no direct influence on the species studied, but it is the source of many indirect impacts modifying the natural mangrove habitat and causing degradation of the latter. The loss and degradation of these habitats severely affect biodiversity and ecological integrity. Mangroves, seagrass beds and coral reefs are closely linked. The goods and services they provide depend on the health of one of these ecosystems. Among other things, the development of infrastructure, breeding ponds, canals and access roads generally leads to changes in local hydrology, salinity and sedimentation.

CHAPTER IV

IV . DISCUSSION

V V.1 SURVEY

Of the 56 locations recorded as likely to have favourable habitats during the individual surveys, 57.1% (n = 32) are home to the Madagascar fish eagle. In the district of Antsalova and in the wetlands of Tambohorano and its surroundings, 84% and 100% of the places indicated during the surveys are home to this species. This reflects the success of The Peregrine Fund's efforts to raise awareness among local populations about the conservation of this species. However, in the 3 sites such as Besalampy, Belo- sur-Tsiribihina and Miandrivazo, local people do not seem to be aware of the value of this species. In fact, the number of habitats in these sites is respectively 40% (n = 2), 50% (n = 5) and 30% (n = 3) of the places recorded during the surveys.

The group survey is interesting despite the fact that the quantitative analysis is not very precise, except in the case of Ankarafantsika National Park. The differences between the results of the group surveys and those of the direct observations could be due to the fact that the people surveyed were not interested in monitoring this species as long as there was no awareness of its conservation, except in the case of Nosy Hara, where the study focused on analysing the vulnerability of marine and coastal avian species to climate change (TPF, 2013). In addition, the group survey in four target PAs in the north-western part of Madagascar (Nosy Hara, Sahamalaza-ile Radama, Ankarafantsika and Sahamalaza) provides an asset for the conservation of the Madagascar fish eagle. Having recognised the importance of *Haliaeetus vociferoides,* the managers of these PAs have taken the initiative to include this species on the list of conservation targets since the 2017 biological year. This conservation measure would make it possible to prevent the spread of this species and could improve the status of the Madagascar fish eagle so that it meets the IUCN criteria of the upper "En" category (critically endangered).

VI .2. SEARCH FOR INDIVIDUALS AND NESTS

Raptors are at the top of the food chain. They typically live in low densities (Filippi-Codaccioni, 2012). These characteristics explain the low rate of encounters during surveys. In addition, the Madagascar fish eagle prefers specific habitats such as wetlands rich in fish and surrounded by tall trees. The systematic observation method combined with surveys therefore seems to be the most effective way of surveying the population of this species.

Some places such as Nosy Androtso, Nosy Tanga, Nosy Sahankazo and Ampasindava located outside the Nosy Hara MPA could host this species because they are indicated during the surveys but are not visited because of the weather constraint. However, other places mentioned in the survey results such as Ampasindava, Andranomavo, Andranoboka, Antsakoa, Nosy Tanga and Rantavorona are not eagle fishing territories but are only occasionally visited by the species during certain periods for hunting. In addition, certain places such as Andranomavo and Rantavorona are heron dormitories.

VII3. POPULATION SIZE

Throughout the Grande Ile, Rabarisoa *et al* (1997) recorded a total of 222 individuals in 63 pairs, while the TPF teams (2005 - 2006) recorded 287 individuals in 121 pairs, 15 wandering adults and seven immatures (Razafimanjato, 2011). In the present study, only a few parts of the species' range were surveyed. Certain sites identified as having a high concentration, namely the Mitsio archipelago, Mahajamba and Moramba bays and other sites scattered in coastal and continental wetlands (Razafimanjato, 2011) are not among the target sites. In fact, 193 individuals have been inventoried, 192 of which are found in the 10 target sites of the last global census and one individual is located in a new site (Lac Tseny). Over the course of a decade, the population at these 10 sites has increased by 24 individuals, representing 12.9% of the population in 2006 (n = 168 individuals). Thus, if the census was carried out throughout the Grande Ile, we could at least find a total of 312 individuals because, at the time of the last global census, the unvisited sites were home to 119 individuals (Razafimanjato, 2011). Moreover, in most of the sites in the present study, an increase in the population has been noted over the last decade.

Based on the numbers of individuals counted by age group, it was found that the Madagascar population of the Fish Eagle has a low number of sub-adult individuals compared to the number of adult and juvenile individuals. This situation could be explained by the species' low reproductive success rate. There is also a high mortality rate among juveniles.

Following the 2005 - 2006 national census, Razafimanjato (2011) found that Sahamalaza Bay was home to more individuals than the complex of three lakes in the Tsimembo Manambolomaty PA, i.e. 29 compared to 26 individuals. This assertion is also proven by the present study, since the numbers of individuals inventoried there are 37 and 36 respectively.

In certain sites such as Antsalova, Sahamalaza, Nosy Hara, Baie de Baly and Ankarafantsika, the abundance of fish eagles is explained by the presence of conservation actions in protected areas. What's more, their ecosystems (coastal or lacustrine) are still in their natural state, with little anthropogenic pressure, allowing the population to increase. On the other hand, the low numbers in river ecosystems could be due, on the one hand, to the disruption of river transport and, on the other hand, to the difficulty this species has in capturing its prey in fast-flowing, turbid rivers.

IV.4.DISTRIBUTION OF WORKFORCE BETWEEN PROTECTED AND UNPROTECTED AREAS

Unprotected wetlands are not protected from human activity (Ratsisompatrarivo *et al.,* 2016). In the case of our study, apart from overfishing, the edges of lakes Tseny and Badika are being cleared due to the extension of rice, groundnut and maize fields. All the lakes visited in the Besalampy district, the Ambalarano and Antevamena lakes near the Mandrozo PA, the Berevo and Andranovoribe lakes in the Belo-sur-Tsiribihina district and the Maroampinga, Amparihy and Borimena lakes in the Miandrivazo district are gradually being transformed into rice fields. This explains the low numbers, and later the disappearance, of this species in unprotected areas. Furthermore, individuals living in these unprotected areas are susceptible to various pressures and threats. This fact is proven by certain cases admitted during the investigations, including the hunting of adult individuals in Lake Amparihy, the collection of eggs and the removal of chicks from their nests in Lake Andranomazava. On the other hand, no traces of hunting were observed and/or reported within the protected areas. In these PAs, the individuals would have a chance of survival because they are the conservation targets.

IV.5. GEOGRAPHICAL DISTRIBUTION OF *Haliaeetus vociferoides*

According to previous studies (Rabarisoa *et al.*, 1999; Razafimanjato, 2011), the geographical distribution of the Madagascar fish eagle is limited to Nosy Hara, in the north, Morombe, in the south, and around Miandrivazo and Ankarafantsika, inland. Currently, the Mangoky Ihotry Complex, which encompasses the wetlands of the Manja and Morombe districts, is no longer home to this species. In fact, the southern limit of the Madagascar fish eagle's geographical distribution area is moving further and further north, as it lies in the Belo-sur-Tsiribihina district, while the northern limit remains unchanged. The disappearance of this species in the southern part could be due, on the one hand, to anthropic disturbance and, on the other, to poor habitat quality. This hypothesis could be justified by the fact observed in Lake Bemamba (which was home to the last known pair in this area): apart from overfishing, this lake was covered with floating vegetation and the edge did not have enough large trees suitable for the nesting of this species.

Furthermore, taking into account climatic factors and forest cover, Razafimanjato's (2011) projection showed that the favourable habitat for the Madagascar fish eagle lies between Antsiranana, in the north, and Morombe, in the south, and widens towards the interior of the mainland, around Ankarafantsika. In addition, the results of his research predict that by 2050 the geographical distribution of the Madagascar fish eagle could extend into the southern and south-western parts of the Grande Ile if conditions are favourable. This will be between the Ambatoboeny district in the north and the Ampanihy district in the south, via the Mangoky river and the Makay plateau. In addition, it anticipated that this river and plateau, as well as the wetlands of Belo-sur-Tsiribihina, would be potential future sites, but a reduction, or even disappearance, of this species in the north-western part would be possible or at high risk. However, the results of this study showed that many lakes in the Belo-sur-Tsiribihina district, such as Mamotsaky, Berevo, Akopoky and Serinamo, no longer harbour this species. According to the findings of Razafimanjato (2011), these lakes have been abandoned over the last decade. And then, until now, the population in the north-western part appears to be increasing. This leads us to ask whether this increase is due to the effort to conserve habitats by setting up PAs or whether this species has a chance of increasing in this area.

Moreover, the geographical distribution of the Madagascar fish eagle allows us to assume that the presence of PAs or conservation actions ensures the survival of this species in its distribution area. This can be explained by the fact that all the areas of occurrence of the Madagascar fish eagle are PAs except for the Mitsio Archipelago and Mahajamba Bay. From north to south, Nosy Hara, Baie de Sahamalaza-ile Radama, Baie de Baly, Ankarafantsika, Mandrozo, Complexe Tsimembo Manabolomaty and Ankevo are all PAs. Mahajamba Bay is not a PA, but aquaculture there has an interest in protecting the mangroves. In fact, up to now, the presence of this project does not run counter to the conservation of the fish eagle in this area. So, to maintain the southern limit of the species' geographical distribution area, it is preferable to create PAs in favourable areas in the Miandrivazo and Belo-sur-Tsiribihina districts.

IV.6. POPULATION TREND

As for the population dynamics of the Madagascar fish eagle, anthropogenic pressures such as hunting, competition for access to fish resources (Strzalkowska *et al.,* 1995), loss of natural habitat (Razafimanjato *et al.,* 2007), the removal of juveniles (Langrand, 1987; Langrand & Goodman, 1995) and natural disasters such as tropical cyclones and high winds (Watson, 1993) are the main causes of population decline.

In the Belo-sur-Tsiribihina district, the population in 2006 (n = 19 individuals) has decreased by 36.8% (n = 7 individuals) over the last decade. This downward trend could be due to the destruction of habitats following the destruction of wetlands, which have continued to increase in this region. During farming activities, most of the lakes are transformed into rice fields. Fish resources are not available in sufficient quantities for local residents and the Madagascar fish eagle population.

In the case of Antsalova, the population of the district as a whole is declining. However, in the GELOSE areas, made up of the complex of three lakes (Ankerika, Soamalipo and Befotaka) and two satellite lakes (Betangiriky and Bekofoky), the population has increased by 6 individuals over this decade, i.e. from 30 (Razafimanjato, 2011) to 36 individuals, including 28 adults, 5 juveniles and 5 sub-adults. The Tsimembo Manambolomaty complex PA, which encompasses the GELOSE area and lakes Andranolava-nord, Antsamaky and Beora, is home to 47 individuals. It would therefore seem that the growth of this species depends on the structure and quality of its living environment.

In addition, we cannot exclude the possibility of an increase in the population of the Madagascar fish eagle in the Nosy Hara, Sahamalaza, Ankarafantsika and Baie de Baly sites as long as the protected areas are well managed. Thanks to the presence of large trees, their tranquillity and the availability of food resources, these sites could still support more individuals.

In all the study sites, population growth of 1.3% per year over the last decade still seems very slow. Thus, the evolution of the situation of the Madagascar fish eagle seems somewhat contradictory, i.e. a reduction in the geographical range but an increase in the total population size. There are three possible reasons for this: (1) the carrying capacity at each site has not yet been reached. This hypothesis is confirmed by the formation of new pairs in most of the sites; (2) the majority of the population counted (63.7%) takes refuge and nests in Protected Areas, i.e. there are management rules; (3) most of the wetlands along the western part of the Grande Ile are classified as PAs. This situation ensures the growth and survival of this species.

IV.7. BREEDING SUCCESS AT THE TWO TPF INTERVENTION SITES

Lewis & Watson (1993) found a value of around 55% of young per nest in their study of 27 pairs in the Antsalova district. This success rate no longer deviates from that found by Razafimanjato (2011) in his study of 128 active nests during the period 1992 - 2007. The present study revealed that this average success has increased by 75% of young per brood for the 214 active nests over the last decade. This rate is comparable to that of the three-lake complex (77% of young per brood) during the GELOSE probationary period (Razafimanjato, 2011). These results show that natural resource management based on the local population has proved effective in the Tsimembo Manambolomaty PA complex.

In the Tambohorano wetland, no information on the breeding success of the Madagascar fish eagle has been reported in the literature. During the present study, the average success rate over the last decade was 61% of young per brood from 35 active nests. This rate is slightly lower than that of the Antsalova district. The difference can be explained, on the one hand, by the relatively low number of samples for the Maintirano district and, on the other hand, by the abundance of food resources available and the security of the nesting territories. For some raptor species, such as the Montagu's Harrier *Circus pygargus* and the Peregrine Falcon *Falco peregrinus*, breeding success varies from year to year with the abundance of food resources (Salamolard *et al.,* 2000) and weather conditions. But it also depends on sites frequented by humans and scrupulous respect for quiet zones.

Compared with the Red Kite *Milvus milvus*, one of only two species of raptor endemic to Europe, the breeding success of the Madagascar fish eagle is low. For the Red Kite, the average nest size (number of young per successful pair) varies annually from 1.63 to 2.33 (Riols, 2014). In the Madagascar fish eagle, juvenile mortality due to cainism or fratricide and attacks by necrophagous insects are the main causes of failure to achieve fledging success (Razafimanjato, 2005). This natural phenomenon (cainism) can be avoided with human intervention by adopting a semi-captive breeding method called "Hacking". The implementation of this technique is considered effective but the cost is unfortunately unaffordable (Razafimanjato, 2011).

IV.8. NIDIFICATION

IV.8.1. Location of nests

Razafindramanana (1995) reported that the nesting trees existing in the complex of three Manamolomaty lakes are located at a maximum distance of 0.29 km from the shore. Razafimanjato (2011) found that this distance reaches up to 0.64 km. He suggested that the strong pressure exerted by fishermen, through non-respect of usage rights such as the harvesting of various forest resources within a 100 m radius of the bank, could be the cause of this inland retreat. In the present study, we found the same distance found by Razafindramanana in 1995. The efforts of The Peregrine Fund to set up the GELOSE project on the conservation of this species, while avoiding the various kinds of disturbance, probably favours this nest location on trees a little closer to the lakeshore. On the other hand, in other lake ecosystems such as Antsamaky and Andranovorinankoay, pairs nest up to 0.7 km from the lakeshore to avoid disturbance.

The nest of the Madagascar fish eagle must be placed very close to a body of water offering sufficient fish (Berkelman *et al.*, 1999; Watson *et al.*, 2000; Razafimanjato, 2011).

In contrast, in the case of the Osprey *Pandion haliaeetus*, the nesting site and fishing sites do not need to be close to each other. This species can cover a maximum distance of 20 km around the nest to collect its prey (Devoulon & Lavarec, 2014). It prefers large wetlands bordered by natural forests or artificial structures close to bodies of water such as fish farms (Dennis, 2016).

IV.8.2. Change of nest location

Like all other birds of prey, the Madagascar fish eagle can build several nests in its territory. Tingay (2005) and Razafimanjato (2011) estimated at least six cases of nest changes during their studies. There are several reasons for nest changes, including early failure of the first nest or last-minute disturbance (Newton, 1979), the availability of large trees and the tranquillity of territories (Razafimanjato, 2010). For the present study, the frequency of passage of fishermen and/or tourists in the vicinity of the nesting site, the ageing of building materials, the felling of nest-bearing trees and the setting on fire of the nesting territory are the main causes of changes in nest location.

IV.8.3. Nest location and characteristics

The Madagascar fish eagle builds its nest with bulky branches placed in a fork at the top of a large tree, on a cliff or on a rocky island (Rabarisoa, 1997). This condition enables the eagle to spot fish (Langrand & Meyburg 1989). In 1999, Watson and his team found that the selection of small sticks varied from 1.5 to 3 cm in diameter and from 30 to 60 cm in length and that the complete nest structure measured around 120 cm, with the log measuring 70 x 50 cm. The results of the present study are comparable with those of Watson and his colleague (1999), since we found an average of 124.25 cm for the complete nest structure. The logis measures 70.75 x 53 cm, and the average external and internal nest depths are 55.5 cm and

13.75 cm respectively (n = 4). From our direct observation, it appears that some nests in the three-lake complex of Tsimembo PA, such as Soamalipo 1, Befotaka 1 and others, are about 1.5 m high. We were unable to measure these nests because they were active during our visit.

IV.8.4. Nest tree characteristics

Razafindramanana (1995) identified that in the complex of three Antsalova lakes (Ankerika, Befotaka and Soamalipo) the nesting trees are dominated by the "Kily" (*Tamaridus indica).* Razafimanjato (2011) noted that the pair of Madagascar fish eagles in the Antsalova district also use the "Baobab" (*Adansonia* sp) for nesting. The present study found that the nest trees identified in the continental zone are dominated by the "Kily" followed by the "Sarongaza" *Colvillea racemosa* and the "Baobab". In addition, we found that these tree species have no particular use in the daily lives of local residents. This is why they are being set aside by illegal felling and removal during the exercise of usage rights.

African fish eagles build their nests even on low trees such as euphorbia (Internet 2). This has also been observed in the Madagascar fish eagle living in coastal areas, where it nests mainly on *Sonneratia alba* or "Farafaka", even at 5 m above ground level. The size of this tree is relatively large compared with other mangrove species.

The height of the nest site varies according to the type of ecosystem. In the case of the present study, the average nest height was 16.42 m (n = 26 nests) in the continental environment, whereas in the littoral zone the average nest height was 9.25 m (n = 14 nests) from the ground. The means of Dhp are 79.31 cm (n = 26 nest trees) and 41.48 cm (n = 14 nest trees) respectively. These differences observed in the dimensions (height and diameter) of the nest trees between the coastal and continental environments could be explained by their difference in type of plant formation. This variation in nest site height is also observed in other raptor species. This is the case for the Granada Crocket-billed Kite *Chondrohirax uncinatus mirus*, whose variation in nest height depends on the type of forest in which the species is found (Thorstrom *et al.*, 2001). On the other hand, the height of *Aviceda cuculoides* nests varies according to the authors: between 20 and 30 m according to Chittenden (1984), 13 m according to Hall *et al.* (1991) and even between 10 and 30 m from the ground (Ferguson-Lees & Christie, 2001).

IV.8.5. Habitat characteristics surrounding the nest

The Madagascar fish eagle needs a habitat with a wide field of vision and therefore a dominant height for nesting (Berkelman, 1997; Berkelman *et al.,* 2002). The choice of nesting territory depends on the number of perches and the size of the tree. A large tree could be a favourable nesting area for this species, and also promotes their safety from egg collectors or eaglets.

Unfortunately, we have noticed that some pairs living in unprotected areas seem to be victims of deforestation. In fact, in these areas there are quite limited areas of the type sought after by the Madagascar fish eagle, particularly in terms of the choice of aquatic environments suitable for fishing and forest areas suitable for nesting. This context explains the reason for the presence of an active nest located no more than 10 m from the camp in the Amparihy wetland during the present study. The owner pair could leave this location due to the lack of habitat quality. The same applies to the Andranovorinankoay pair, as their nest is built on a tree located inside the plot of land that has been burnt by bushfire.

In the coastal zone, the mangrove is currently under heavy pressure from human activities. Apart from local use, wood is trafficked to towns such as Mahajanga (Razafimanjato, 2011).

In our case, we noticed numerous movements of mangrove wood: from Belo-sur-mer and Belo-sur-Tsiribihina to Morondava, from Tambohorano to Maintirano, from Masoarivo to Antsalova, from Besalampy and Soalala to Mahajanga. In addition, the development of shrimp farming has led to the degradation of habitats associated with mangroves, such as wetlands, salt marshes and lagoons. The loss of these habitats and the increase in pollution associated with shrimp farming are also threatening adjacent marine ecosystems such as coral reefs and seagrass beds (Francoeur, 2009).

Concerning the distance between two neighbouring nests, for the three Antsalova lakes complex, Watson *et al.* (1993) found an average of 1.48 km, Razafimanjato (2011) noted that in 2005 this distance was approximately 1.98 km. In the present study, the average distance between two nests differed by 1.84 km (n = 10 nests). This difference could be due to the change in nest location made by existing pairs over time. Indeed, during his study between 1992 and 2007, Razafimanjato (2011) confirmed that a pair can move their nest around six times in order to reproduce.

The satellite lakes, as their name suggests, are scattered throughout the Antsalova district. The previous study found that the 17 nests found were 3.27 km apart (Razafimanjato, 2011). In contrast, during the present study, the average distance between two neighbouring nests in satellite lakes in the Antsalova district increased by 9.17 km (n = 10 nests). This spacing can be explained not only by changes in nest location but also by the disappearance of seven nests. Furthermore, during this study, no lake hosted more than one breeding pair, with the exception of Ankerika, Befotaka and Soamalipo. The territorial behaviour of this species would not allow at least two breeding pairs to be established in the other lakes. It is for this reason that the parents chase the young eagles to settle elsewhere. On the other hand, in the case of *Pandion haliaetus*, a species with the same habitat preference and food preferences as the Madagascar fish eagle, the establishment of a pair in a particular location is influenced by behavioural factors, in particular the phenomena of philopatry and intraspecific attraction. In 85% of cases, the young born in a location will return to the same location to reproduce. In addition, the presence of established pairs attracts new individuals (Devoulon & Lavarec, 2014).

In coastal areas, Razafimanjato (2011) found that nests are spaced an average of 6.10 km apart (n = 34 nests). This distance is close to that of Nosy Hara in 2016, where it was 6.62 km (n = 7 nests). However, it is well above the average found for the 8 nests located at Sahamalaza (5.24 km). Coastal areas are available for nesting Madagascar fish eagles. However, disturbance caused by the high number of fishermen, deforestation, depletion of fish resources and poor water quality are factors that may explain the absence of pairs in unoccupied areas. According to Razafimanjato (2011), these various causes create unfavourable conditions for the sedentarisation and nesting of the Madagascar fish eagle.

Moreover, the absence of the nests of other raptor species around the nest of the Madagascar fish eagle is probably due to the latter's territorial defence behaviour. Each pair defends their home range in the presence of intruders. This territorial defence could end in a bloody fight (Razafimanjato, 2010).

CONCLUSION AND RECOMMENDATIONS

The study of the Madagascar fish eagle *Haliaeetus vociferoides* carried out at 11 sites scattered across the western part of the Grande Ile, between Nosy Hara in the north and Morombe in the south, provides up-to-date information on the population size, distribution and habitat characteristics of the species in question. Data on pressures and threats to the species and its natural habitat were also presented in this study. The information contained in this work serves as an effective tool for site managers and promoters in developing conservation strategies.

The results of the census revealed a relatively slow increase in the remaining population of the Madagascar fish eagle, of the order of 1.3% individuals/year, for all the sites visited over the last decade. We therefore estimate that the overall population will continue to grow, but slowly. This hypothesis is confirmed if we add to the results of the present study the numbers recorded at the sites not visited, i.e. almost 119 individuals. We can therefore conclude that Madagascar is probably home to a population of around 312 individuals at present. This means that the population has increased over the last decade, from 287 to 319 individuals. This increase augurs well for the conservation of the species in Madagascar and for biodiversity in general.

As for the distribution area, it is limited only between Nosy Hara and Belo-sur- Tsiribihina, synonymous with a decrease in distance and compression of surface area on the western part, as previous studies have mentioned that it extends as far as the Morombe district. In addition, an extension as far as Maevatanana was recorded for the inland limit, instead of PN Ankarafantsika previously. As a result, the hypothesis stipulating the immutability of the distribution limits of the species in question is not verified by this study.

As far as ecology is concerned, the Madagascar fish eagle builds its nest on large trees with a diameter varying from 26 to 186 cm and a height of 5 to 24 m, all species combined (mangroves, dry forest).

In addition, during the course of this study, it can be concluded that the survey method appears to be effective for monitoring the Madagascar fish eagle population and not for collecting information on its reproduction. In general, local residents are only interested in the presence or absence of the species in a given environment. Other information is excluded from their daily observations.

A number of recommendations are put forward to strengthen conservation measures for the Madagascar fish eagle.

It appears that the census of the population covering the range of the population of this species was not carried out during the present study. In order to have more information on the current size of its overall population, it is necessary to continue the census in the areas not visited during this study.

To better protect and conserve this species in their specific habitats, human activities leading to the destruction of wetlands must be minimised. The aim of this recommendation is to better ensure the availability of food and nesting sites. To achieve this, villagers and fishermen need to be made aware of the importance of wetlands and their surrounding forests for the survival of the Madagascar fish eagle and other related species. In addition, strengthening fisheries and forestry surveillance as well as the capacity of the local population in terms of natural resource management is becoming an essential necessity in order to contribute to improving the reproduction of the Madagascar fish eagle and possibly other species of waterfowl. In

addition, it is essential to seek alternative ways of increasing the income of local people who depend on wetlands.
Action should be taken at sites where this species is not yet a conservation target, by visiting nests at least three times a year, especially in areas of high concentration. This will make it possible to assess the productivity of this species in these areas. To achieve this, a National Action Plan (NAP) for the Madagascar fish eagle needs to be drawn up in collaboration with PA managers and/or the local population, and the knowledge of grassroots communities in terms of participatory ecological monitoring (PEM) needs to be strengthened.
Given their habitat requirements, the resilience of the Mada Gascar fish eagle could be increased. To achieve this, in addition to the sustainable management of fishing activities, it would be necessary to conserve the continental wetlands with their surrounding forests, as well as the mangroves in the coastal zone. In addition, reforestation and the dispersal of fry in degraded wetlands, such as lakes, would be feasible.
It would be interesting to improve the system of environmental education in schools.

BIBLIOGRAPHICAL REFERENCE

1. **Andreianov, B., Caldini C., Donadello C. & Klein P., 2011:** Mathématiques pour la modélisation Dynamique des populations. University of Franche-Comté. 44p.

2. **Andreone, F., Vences, M., Glos, J. & Glaw F., 2010:** Molecular and bioacoustics differentiation of *Boophis occidentalis* with description of a new tree frog from northwestern Madagascar Magnolia Press, *Zootaxa 2544.* Pp: 54 - 68.

3. **Andriafidison, D., Richard K. B. J., Paul V. L., Caskie T. Mc., Andrinajoro A. R., Rahalambomanana, J., Ravelomanana, T., Raminosoa, N. & Saunders, A., 2011:** Preliminary fish survey of Lac Tseny in northwestern Madagascar. *Madagascar Conservation &Development*, Vol 6, No 2 (2011). Available at http://www.joumalmcd.com/index.php/mcd/rt/printer Friendly/mcd.v6i2. 7/0.

4. **ANGAP, 2002:** Plan de gestion, Parc National de Baie de Baly. 69p.

5. **Berkelman, J., 1997:** Habitat requirements and foraging ecology of the Madagascar FishEagle. Doctorate of Philosophy in Fisheries and Wildlife Sciences. Virginia Polytechnic and State University. Blacksburg, Virginia. USA. 137p.

6. **Berkelman, J., Fraser, J.D. & Watson, R. T., 2002:** Nesting and perching habitat use of the Madagascar fish eagle. *Journal of Raptor Research* 36(4): 287-293.

7. **Birdlife International, 2010:** *Threatened birds of the world 2010.* CD-ROM. Cambridge, UK: Birdlife International.

8. **Boiteau, P., 1999:** *Dictionnaire des noms malgaches des végétaux.* Boiteau, P., Boiteau, M. & Allorge-Boiteau, L. (eds). Edition Alzieu. 4 volumes.

9. **Chittenden, H., 1984:** Aspects of Cuckoo Hawk *Aviceda cuculoides* breeding biology. In *Proceedings of the second symposium on African predatory birds*. Mendelsohn, J. M. & Sapsford, C. W. (eds.). Natal Birds Club, Durban. Pp. 47-56.

10. **Clauvice, A.B., 2014:** Valeur de Gestion de la mangrove dans le Parc National Sahamalaza-Ile Radama: cas de la parcelle Ampasimbezo. Final dissertation in view of obtaining the degree of Licence Professionnelle Es-Science. Faculty of Science, Technology and the Environment, University of Mahajanga. 46p.

11. **Collar, N.J. & Stuart, S.N., 1985:** Threatened Birds of Africa and Related Islands. International Council for Bird Preservation/International Union for the Conservation of Nature, Cambridge.

12. **Collar, N.J.; Crosby, M. J. & Statterfield, A. J., 1994:** Birds to watch 2. The world list threatened birds. *BirdLife International*, Cambridge.

13. **Dennis, R., 2016:** Recovery and safeguard plan for the osprey in Europe, particularly in the Mediterranean basin, Strasbourg. 23p.

14. **Devoulon, A. & Lavarec, L., 2014:** Synthèse 2013 - 2014 sur la situation du Balbuzard pêcheur en Ile-de-France. 29p.

15. **Durell, 2014:** Etudes d'impacts environnementaux et sociaux relative à la mise en place de la nouvelle Aire Protegee Menabe Antimena région Menabe. 123p

16. **Ferguson-Lees, J. & Christie, D. A., 2001:** Raptors of the world. Houghton Mifflin Company, Boston. 992 pp.

17. **Filippi-Codaccioni, O., 2012:** The impact of climate change on bird migration in Aquitaine *Post-doctoral research report Version 1.0.* 63p.

18. **Francoeur M., 2009:** L'élevage de la crevette : une menace pour les mangroves? Master's thesis in International Ecology Department of Biology Faculty of Science University

of Sherbrooke, Quebec, Canada, 92p.

19. **Fowler, R. A. & Cohen, L., 1985:** *Statistics for ornithologists*. British Trust for Ornithology, London. 175p.

20. **Fuchs, J., J.M. Pons, E. Pasquet, M.J. Raherilalao & Goodman, S. M., 2007:** Geographical structure of genetic variation in the Malagasy Scops-Owl inferred from mitochondrial sequence data. The Condor 109. Pp: 408 - 418.

21. **Glaw F, Kohler J., Townsend T.M. & Vences, M., 2012:** Rivaling the World's Smallest Reptiles: Discovery of Miniaturized and Microendemic New Species of Leaf Chameleons (Brookesia) from Northern Madagascar. University of Lausanne, Switzerland February 2012. 24p.

22. **Goodman, S. M. & Hawkins, A. F. A., 2008:** Birds. In *Paysages naturels et biodiversité de Madagascar*. Publications scientifiques du Museum, Paris, WWF. Pp: 383 - 435.

23. **Hall, D. G., Tarboton, W. R. & Theron, F., 1991:** More on the enigmatic Cuckoo Hawk. Gabar, 6. pp: 47 - 50.

24. **Humbert. H. & Leandri. J. 1954:** Cinquante ans de recherché botanique à Madagascar, *Bull, Acd. Malg*. special fiftieth anniversary. Pp: 33 - 42

25. **Jonhson, R.R., 1992:** *Elementary Statistics*. Sixth edition. PWS-KENT, USA. 733p.

26. **Koechlin. J. & Guillaumet. J. L., 1974:** Flore et végétation de Madagascar, Vaduz, Cramer ed. 687p.

27. **Langrand, O., 1987:** Distribution, status and conservation of the Madagascar fish-eagle *Haliaeetus vociferoides* Desmurs 1845. *Biological Conservation*, 42: 73-77.

28. **Langrand, O., 1995:** *Guide des Oiseaux de Madagascar*. Yale University Press, New Haven & London. 415p.

29. **Langrand, O. & Goodman, S., 1995:** Monitoring Madagascar's ecosystems: a look at the past, present, and future of its wetlands. Pp: 204 - 214. In *Ecosystem monitoring and protected areas: Proceedings of the Second International Conference on Science and Management of Protected Areas, 16 - 20 May 1994, Nova Scotia*, eds. T. B. Herman, S. Bondrup-Nielson, J. H. M. Willison and N. W. P. Munro. Nova Scotia Science and Management of Protected Areas Association.

30. **Langrand, O. & Meyburg, B.-U., 1984:** Birds of prey and owls in Madagascar: their distribution, status and conservation. Pp: 3 - 13. in *Proceedings of the Second Symposium on African Predatory Birds, 22 - 26 August 1983, Durban, South Africa*, eds. J.M. Mendelsohn and C.W. Sapsford. Natal Bird Club.

31. **Langrand, O. & Meyburg, B.-U., 1989:** Range, status and biology of the Madagascar fish-eagle (*Haliaeetus vociferoides*). Pp: 269 - 278. *in Raptors in the Modern World: Proceedings of the III World Conference on Birds of Prey and Owls, 22 - 27 March 1987, Eilat, Israel*, eds. B.U. Meyburg and R. D. Chancellor. World Working Group on Birds of Prey and Owls, London.

32. **Léo M. & Bertin. A., 2012:** *Understanding, constructing and interpreting statistics.* Version of 30 March 2012. 16p.

33. **MEEF, 2012:** Madagascar: State of the Environment Report 2007 - 2012. Available at http://mg.chmcbd.net/implementation/ Documents nationauxnationaux/rapport sur l'état de l'environnement - 2012.

34. **Meyburg, B.-U., 1986:** Threatened and near-threatened birds of prey of the world. *Birds of Prey Bulletin* 3: 1-12.

35. **MNP, 2016:** Conservation of Angonoka in the Baie de Baly National Park. Madagascar

National Parks fights Angonoka trafficking in the Baie de Baly Park. Antananarivo Madagascar. 3p.

36. **Morris, P. & Hawkins, A.F. A., 1998:** Birds of Madagascar. A Photographic Guide. Pica Press. 316p.

37. **Oberlé, P., 1981:** Madagascar: un sanctuaire de la nature. Librairie de Madagascar. 115p.

38. **Pruvot, Y.Z.M., 2017:** Bioecologique et abondance du Râle d'Olivier *Amaurornis olivieri* dans l'Aire Protégée Madrozo, district de Maitirano Région Melaky. Mémoire en vue de l'obtention du Diplôme d'Etudes Approfondies, Université de Toliara. 72p.

39. **Rabarisoa, R., 1999:** Nouveaux éléments sur la distribution de Pygargue de Madagascar *Haliaeetus vociferoides* dans la région du sud-ouest de Madagascar. Pp: 110114. In *Projet de Conservation des Zones Humides de Madagascar: Rapport d'Avancement V, 1997 &* 1998. Eds. A. Andrianarimisa. The Peregrine Fund, Antananarivo, Madagascar.

40. **Rabarisoa, R., 2000:** Madagascar Fish Eagle Offshore survey in the West and Northwestern Coast of Madagascar. Unpublished Report, The Peregrine Fund, Madagascar.

41. **Rabarisoa, R., Watson, R. T., Thorstrom, R. & Berkelman, J., 1996:** Status of the Madagascar Fish Eagle *Haliaeetus vociferoides*. In *Madagascar Wetlands, Conservation Project. Progress report III, 1995-1996*, eds. R.T. Watson. The Peregrine Fund, Boise.4 16p.

42. **Rabarisoa, R., Watson, R. T., Thorstrom, R. & Berkelman, J., 1997:** Status of the Madagascar Fish Eagle *Haliaeetus vociferoides* in 1995. *Ostrich* 68(1): 8 - 12.

43. **Rabearivony, J., Thorstrom, R., Rene de Roland, L. A., Rakotondratsima, M., Andriamalala T. R. A., Sam, T. S., Razafimanjato, G., Rakotondravony, D., Raselimanana, A. P. & Rakotoson, M., 2010:** Protected area surface extension in Madagascar: Do endemism and threatened species remain useful criteria for site selection? *Madagascar Conservation & Development*, Volume 5, Issue 1: 35 - 47.

44. **Rabenandrasana M., Zefania S., Long P., Sam, T. S., Virginie M. C., Randrianarisoa M., Safford R. & Székely T., 2009:** Distribution, habitat and status of "Endangered" Sakalava Rail of Madagascar. *Bird Conservation International* 19(1): 23-32.

45. **Rafanomezantsoa, S. A., 1998:** Contribution des comportements et de la dispersion des jeunes Pygargues de Madagascar *Haliaeetus vociferoides* (Desmurs, 1845) dans le complexe des trois lacs d'Antsalova. Mémoire de Diplôme d'Etudes Approfondies, Université d'Antananarivo, Madagascar. 70p.

46. **Rakoto, N. J. P., 2005:** Contribution à l'amélioration de la production des semences riz dans la région Sud-Ouest Malgache. Diplôme d'Etudes Approfondies dissertation, University of Toliara. 68p.

47. **Rakotomalala, R., 2011:** Tests de normalité. *Empirical techniques and statistical tests* Version 2.0 Université Lumière Lyon 2. Pp 53.

48. **Ralison, H. O., Rakotondrazafy, H. H., Leone, M. & Rakoto Ratsimba, H., 2011:** Marine Protected Areas and Climate Change - Experiences from Nosy Hara Marine National Park. WWF. 36p.

49. **Rasamoelina, A.D., 2000:** Contribution à l'étude de la faune ichtyologique du complexe des trois lacs : Befotaka, Soamalipo et Ankerika dans la région d'Antsalova. Mémoire de Diplôme d'Etudes Approfondies, Université d'Antananarivo, Madagascar. 70p.

50. **Raselimanana, A. P., Raherilalao, M. J., Soarimalala, V., Gardner, C. J., Jasper, L. D., Schoeman, M. C. & Goodman, S. M., 2012:** An initial overview of the vertebrate fauna of the Salary-Bekodoy spiny bush, west of Mikea National Park, Madagascar.

MalagasyNature,

6: 1-23.

51. **Ratsisompatrarivo, J. S. & Rasoamampianina, V. A., 2016:** Biodiversity Conservation and Poverty Reduction in Madagascar. Network of Conservation Educators and Practitioners, Center for Biodiversity and Conservation, American Museum of Natural History. *Lessons in Conservation*, Vol. 6: 30-61.

52. **Raveloson, L., 2017:** Étude biologique et écologique du Faucon a ventre raye *Falco zinoventris*, dans l'Aire Protégée Complexe Tsimembo Manambolomaty, partie Ouest de Madagascar. Dissertation for the Diplôme d'Etude Approfondies. Université de Toliara.72p.

53. **Razafimanjato, G., 2005:** Productivité et mortalité de Pygargue de Madagascar *Haliaeetus vociferoides* avant et après mise en place de la GELOSE dans le complexe de trois lacs : Soamalipo, Befotaka et Ankerika dans la région d'Antsalova. In *Forum de la recherche "Recherche performante: secteur prioritaire, moteur du développement durable"*. Eds. Ministry of National Education and Scientific Research. Pp 13.

54. **Razafimanjato, G., 2010:** Ankoay : des efforts fructueux pour sa sauvegarde. In *Bulletin Trimestriel Songadina n°7*. Conservation International. Pp 3-8.

55. **Razafimanjato, G., 2011:** Conservation de l'aigle pêcheur de Madagascar *Haliaeetus vociferoides* Desmurs, 1845 : Etude bio-écologique et impacts de l'application de la "GELOSE". PhD thesis in Animal Biology, Ecology and Conservation, Department of Animal Biology, Faculty of Sciences, University of Antananarivo, Madagascar. 177p.

56. **Razafimanjato, G., Sam, T.S., Rakotondratsima, M.P.H., Rene de Roland, L-A. & Thorstrom, R., 2014:** Population status of the Madagascar Fish Eagle in 2005-2006. *Bird Conservation International*, 24(1): 88-99.

57. **Razafimanjato, G., Sam T.S. & Thorstrom R., 2007:** *Waterbird Monitoring, in the Antsalova region, Western Madagascar*. Waterbirds, 30(3): 441 - 447.

58. **Razafindramanana, S., 1995:** *Contribution à l'étude de la Biologie de Haliaeetus vociferoides, Desmurs (Pygargue de Madagascar): reproduction et domaine vital*. Mémoire de Diplôme d'Etudes Approfondies, Université d'Antananarivo, Madagascar. 78p.

59. **Razaiarimanana, J., 2010:** Approches de la conservation durable de *Dioscorea maciba* au Parc National Ankarafantsika (Nord-ouest de Madagascar). In *Les ignames malgaches, une ressource à préserver et à valoriser. Proceedings of the symposium held in Toliara, Madagascar,* 29 - 31 July 2009. Eds. Tostain S., Rejo-Fienena F. Pp. 6-11.

60. **Riols R., 2014:** Bilan du programme d'étude et de conservation du Milan Royal en Auvergne (période 2005-2013). LPO Auvergne. 52p

61. **Salamolard, M., Butet, A., Leroux, A. & Bretagnolle, V., 2000:** Responses of an avian predator to variations in prey density at a temperate latitude. *Ecology,* 81: 2428-2441.

62. **Strzalkowska, S., Kalavah, L. & Mampiandra, J., 1995:** Village dialogue: gaining information through participatory discussions. Pp: 81-132. In *Madagascar Project, Wetlands Conservation Project, Progress Report II, 1993 & 1994*. Watson, R.T. The Peregrine Fund, Boise, Idaho.

63. **Tingay, R. E., 2000:** Sex, Lies and Dominance: Paternity and Behaviour of Extra-pair Madagascar Fish-Eagle *Haliaeetus vociferoides*. MSc thesis, University of Nottingham, UK. 76p.

64. **Tingay, R. E., 2005:** Historical distribution, contemporary status and cooperative breeding in the Madagascar Fish Eagle: Implications for conservation. Dissertation submitted for the degree of Doctor of Philosophy, University of Nottingham, UK. 390p.

65. **Tingay, R.E., Culver, M., Hallerman, E.M., Watson, R.T. & Fraser, J.D., 2002:**

Subordinate males sire offspring in Madagascar fish eagle (*Haliaeetus vociferoides*) polyandrous breeding groups. *Journal of Raptor Research* 36(4): 280-286.
66. **The Peregrine Fund, 2013a:** Analysis of the vulnerability of marine and coastal avian species in Nosy Hara MPA to climate change. 66p.
67. **The Peregrine Fund, 2013b:** Plan d'Aménagement et de Gestion de la Nouvelle Aire Protégée de Mandrozo. 77p.
68. **Thomas, 2005:** Components of the special distribution of a predator: respective effects of habitat, food resources and behavioural interactions. Analysis of non-homogeneous point processes. PhD thesis, Université Claude Bernard Lyon. 115p.
69. **Thorstrom, R., Massiah, E. & Hall, C., 2001:** Nesting biology, distribution and population estimate of the Grenada Hook-billed Kite *Chondrohierax uncinatus mirus*. *Caribbean Journal of Science*, 37: 278-281.
70. **Village, A., 1990:** Raptors biology: Reproduction. In *Birds of Prey*. Newton, I. (ed), Fact on File, Inc, U.S.A. Pp: 124-139.
71. **IUCN, 2015:** *Red List of Threatened Species*. <www.iucnredlist.org>.
72. **Velpry, L., 2015:** World population trends and average annual growth rates. Introduction to demography. 14p.
73. **Watson, R.T., 1993:** (Ed.). Madagascar Project, Progress Report I, 1991 & 1992. The Peregrine Fund, Boise, Idaho, USA.
74. **Watson, R.T., Andrianarimisa, A., Razandrizanakanirina, D. & Kalavah, L., 2000:** Madagascar Fish-Eagle *Haliaeetus vociferoides* and Community-based Wetland Conservation. Pp: 333-342. In *Raptors at Risk*. Chancellors Eds. World Working Group for Birds of Prey/Hancock House.
75. **Watson, R.T., Berkelman, J., Lewis, R. & Razafindramanana, S., 1993:** Conservation studies on the Madagascar fish eagle (*Haliaeetus vociferoides*). *Proceedings of the Pan-African Ornithological Congress* 8: 192-196.
76. **Watson, R.T., Razafindramanana, S., Thorstrom, R. & Rafanomezantsoa, S. A., 1999:** Breeding biology, extra-pair birds, productivity, siblicide and conservation of the Madagascar fish eagle. *Ostrich* 70. Pp: 105-111.
77. **Whitney, D.R. & Mann, H. B. 1947:** On at test of whether one of two random variables is stochastically larger than the other. *Annals of Mathematical Statistics*. 18(1): 50-60.
78. **Young, H.G., 1995:** The Madagascar teal, a most enigmatic duck. *Bulletin African Bird Club* 2(2): 98-100.
79. **ZICOMA, 1999:** Les Zones d'Importance pour la Conservation des Oiseaux à Madagascar. ZICOMA project, Antananarivo, Madagascar. 266p.

Internet links

1. **AmericanOrnithologists ' Union (AOU), 2010** [Online online] http://www.globalraptors.org/grin/site Links.asp?lid=1002. Accessed 18 June 2016.
2. **Internet 1:** https://www.Infoanimales .com/ la-tortuga-angonoka. Accessed 27 June 2016.
3. **Internet 2:** http://www.oiseaux.net/oiseaux/pygargue.vocifere.html. Accessed on 26 November 2017.
4. **The IUCN Red List of Threatened Species. Version 2017-3**. <www.iucnredlist.org>. Accessed 19 January 2018.

APPENDICES

Sites	Place visited	Latitude	Longitude	Ecosystem	Status	Average travel
AP Nosy Hara and surrounding area	Nosy Nosy Hara	12°15'10,2"	49°00'08,6"	Coastline	P	Vedette
	Anjombavola	12°15'56,7"	48°58'33,4"	Coastline	P	Vedette
	Lakandava	12°15'34,9"	48°57'22,8"	Coastline	P	Vedette
	Rantavorona	12°18'05,8"	49°05'09,5"	Coastline	P	Vedette
	Andranomavo	12°18'52,7"	49°03'51,7"	Coastal	P	Vedette
	Anjiamaloto	12°30'40,3"	48°53'47,1"	Coastline	P	Vedette
	Andranoboka	12°20'08,5"	48°59'36,6"	Coastline	P	Vedette
	Ankonkofaly	12°23'39,3"	48°45'42,8"	Coastline	P	Vedette
	Andovobe	12°24'25,4"	48°46'41,9"	Coastline	P	Vedette
	Fararano	12°24'59,8"	48°51'04,2"	Coastline	P	Vedette
	Nosy Valiha	12°22'55,2"	48°41'48,6"	Coastline	NP	Vedette
	Mahanonoka	12°26'56,8"	48'39'52,5"	Coastline	NP	Vedette
AP Sahamalaza-Radama Island and surrounding area	Ankijanimanjofo	14°16'39,1"	48°01'14,9"	Coastline	P	Vedette
	Anjiamalandy	14°05'11,2"	48°01'47,3"	Coastline	P	Vedette
	Vavanankaramihely	14°06'47,3"	48°02'42,0"	Coastline	P	Vedette
	Ankerambalavo	14°04'12,5"	47°58'44,3"	Coastline	P	Vedette
	Ambavanambatomadoso	14°07'38,5"	48°02'23,1"	Coastline	P	Vedette
	Vavanambodidimaka	14°10'28,0"	47°56'45,7"	Coastline	P	Vedette
	Ambohikiriny	14°10'04,0"	47°56'53,4"	Coastline	P	Vedette
	Antranonkira	14°14'04,0"	47°56'08,9"	Coastline	P	Vedette
	Ankitsika	14°17'31,69"	47°59'30,60"	Coastline	P	Vedette
	Nosy Kalanoro	14°11'13,46"	48°01'35,85"	Coastline	P	Vedette
	Sijoro	14°07'50,6"	48°57'12,7"	Coastline	P	Vedette
	Ankorobe	14°07'57,7"	47°59'49,3"	Coastline	P	Vedette
	Ambahofaho	13°57'04,6"	47°58'49,9"	Coastline	NP	Vedette
	Andrafia	13°55'49,5"	47°57'03,1"	Coastline	NP	Vedette
Lake Tseny	Lake Tseny	15°40'41,8"	47°50'33,1"	Lake	P	Foot and canoe
AP Ankarafantsika	Ravelobe	16°18'07,3"	46°49'11,0"	Lake	P	Foot and canoe
	Doanibe	16°20'34,4"	46°44'18,5"	Lake	P	Foot and canoe
	Tsimaloto	16°13'24,1"	47°08'14,8"	Lake	P	Foot and canoe
	Komandria	16°20'49,9"	46°39'23,1"	Lake	P	Foot and canoe
AP Baie de Baly	Ankarananjy	16°01'31,4"	45°30'08,2"	Coastline	P	Vedette
	Sanda	16°01'18,7"	45°24'24,2"	Coastline	P	Vedette
	Lake Sarika	15°58'42,4"	45°11'59,8"	Lake	P	Vedette
	Ambatomainty	16°04'40,8"	45°05'23,2"	Coastline	P	Vedette
	Ankoro	16°09'21,3"	45°10'58,9"	Lake	P	Foot
	Ankoro	16°09'43,1"	45°10'30,1"	Lake	P	Foot
	Kapoloza	16°09'16,3"	45°03'47,9"	Coastline	P	Vedette

Sites	Target location	Latitude	Longitude	Ecosystems	Status	Average travel
District of Besalampy	Amparihy	16°41'36,1"	44°49'38,0"	Lake	NP	Motorbike and pirogue
	Maintimaso	16°38'25,2"	44°45'46,2"	Lake	NP	Motorbike and pirogue
	Ampandra	16°39'33,9"	44°49'18,6"	Lake	NP	Motorbike and pirogue
	Sahapy	16°31'50,7"	44°42'31,7"	Lake	NP	Motorbike and pirogue
	Ankinga	16°33'00,4"	44°27'38,2"	Coastline	NP	Motorbike
Maintirano District	Ambalarano	17°08'05,6"	44°14'41,8"	Lake	NP	Motorbike
	Bezozoro	17°07'17,3"	44°14'39,8"	Lake	NP	Motorbike

	Andranonakoay	17°12'57,0"	44°16'00,9"	Lake	NP	Motorbike
	Mahiarere	17°04'08,2"	44°08'02,7 "	Lake	P	Pirogue
	Nosy Sarotra	17°12'57,0"	44°16'00,9"	Lake	P	Pirogue
	Antevamena	17°36'56,2"	44°10'14,1"	Lake	NP	Pirogue
	Tambohorano	17°28'14,3"	43°56'57,2"	Coastline	P	Pirogue
Antsalova district	Befotaka 1	19°01'36,6"	44°23'58,8"	Lake	P	Pirogue
	Befotaka 2	19°00'47,9"	44°23'46,4"	Lake	P	Pirogue
	Befotaka 3	19°00'25,9"	44°24'36,9"	Lake	P	Pirogue
	Somalipo 1	19°01'17,5"	44°25'09,9"	Lake	P	Pirogue
	Somalipo 2	18°59"09,8"	44°25'18,5"	Lake	P	Pirogue
	Somalipo 3	19°01'29,9"	44°25'58,8"	Lake	P	Pirogue
	Ankerika 2	19°00'32,0"	44°26'59,3"	Lake	P	Pirogue
	Ankerika 3	18°59'09,7"	44°25'18,6"	Lake	P	Pirogue
	Ankerika 4	19°00'32,5"	44°26'59,0"	Lake	P	Pirogue
	Ankerika 5	19°00'32,0"	44°26'59,3"	Lake	P	Pirogue
	Soahany	18°42'02,2"	44°31'49,1"	River	NP	Pirogue
	Androibe	18°42'02,2"	44°31'49,1"	River	NP	Motorbike
	Antsahafa	18°48'42,2"	44°28'38,1"	Lake	NP	Motorbike
	Ambohibary	18°48'28,9"	44°24'48,5"	Lake	NP	Motorbike
	Masama	18°51'01,9"	44°27'16,4"	Lake	NP	Motorbike
	Beora	18°52'43,8"	44°22'03,4"	Lake	P	Motorbike
	Andranolava-nord	19°00'25,9"	44°20'56,4"	Lake	P	Motorbike
	Andranolava-sud	19°05'27,5"	44°24'43,1"	Lake	NP	Motorbike
	Antsamaky	19°02'52,3"	44°21'43,1"	Lake	P	Motorbike
	Antsakotsako	19°06'10,5"	44°32'42,4"	Lake	NP	Motorbike
	Antsohale	19°07'41,4"	44°35'05,3"	Lake	NP	Motorbike
	Ampozabe	19°08'41,9"	44°39'49,5"	Lake	NP	Motorbike
	Bekopaka	19°08'23,3"	44°48'38,2"	River	NP	Motorbike

Sites	**Target location**	**Latitude**	**Longitude**	**Ecosystems**	**Status**	**Average travel**
	Tsinjorano	19°10'22,7"	44°43'44,0"	Lake	NP	Motorbike
	Andranovoribe	19°12'11,6"	44°50'28,2"	Lake	NP	Motorbike
	Maromahia	19°12'21,2"	44°37'10,1"	Lake	NP	Motorbike
	Bejijo	19°13'29,2"	44°32'12,4"	Lake	NP	Motorbike
	Bevahomaty	19°17'31,2"	44°43'57,9"	Lake	NP	Motorbike
	Mamotsaky	19°44'48,2"	44°51'28,6"	Lake	NP	Motorbike
	Antsalaza	18°52'49,1"	44°34'51,2"	Lake	NP	Motorbike
Belo-sur-Tsirbihina district	Ankevo	19°12'30,6"	44°31'52,3"	Lake	P	Motorbike
	Ambatomainty	19°44'25,0"	44°55'20,0"	River	NP	Motorbike and pirogue
	Akimanomby	19°46'53,7"	44°43'51,7"	Lake	P	Motorbike and pirogue
	Berevo	19°41'33,0"	45°03'47,4"	Lake	NP	Motorbike and pirogue
	Akopoky	19°42'46,7"	44°51'16,0"	Lake	NP	Motorbike and pirogue
	Mamotsaky	19°44'50,2"	44°51'28,6"	Lake	NP	Motorbike and pirogue
	Serinamo	19°41'37,2"	44°41'29,6"	Lake	NP	Motorbike and pirogue
	Soarano	19°44'59,9"	44°24'22,9"	Coastline	P	Pirogue
	Nosim-borona	19°49'02,5"	44°25'02,5	Coastline	P	Pirogue
	Anketrevo	19°44'13,4"	45°18'27,0"	Lake	NP	Motorbike and pirogue
	Andranomazava	19°43'26,1"	45°18'52,6"	Lake	NP	Motorbike and pirogue
	Maroampinga	19°40'20,2"	45°17'57,2"	Lake	NP	Motorbike and pirogue
District Miandrivazo	Ankazomanga	19°39'12,0"	45°19'24,6"	Lake	NP	Motorbike and pirogue
	Borimena	19°41'20,4"	45°21'23,7"	Lake	NP	Motorbike and pirogue
	Ampanihy	19°29'05,8"	45°24'04,0"	Lake	NP	Motorbike and pirogue
	Asonjy	19°50'36,5"	45°26'30,9"	Lake	NP	Motorbike and pirogue
	Tsimalaijohary	19°46'05,8"	45°30'18,6"	Lake	NP	Motorbike and pirogue

	Badika	20°09'14,3"	45°31'51,9"	Lake	NP	Motorbike and pirogue
	Anjanaborona	19°48'51,6"	45°34'16,3"	River	NP	Motorbike and foot
Belo-sur-mer to Morombe	Amboboky	20°46'23,6"	43°58'00,4"	Coastline	P	Pirogue
	Marohita	20°45'51,8"	43°59'61,6"	Coastline	P	Pirogue
	Andranopasy	21°17'14,5"	43°42'41,2"	Coastline	P	Pirogue
	Mangoky	21°42'09,7"	43°46'30,5"	River	P	Motorbike and pirogue
	Morombe	21°45'17,3"	43°20'52,1"	Coastline	P	Motorbike

APPENDIX 2: IMAGES SHOWING THE TARGET LOCATIONS FOR THE OBSERVATIONS

> **Nosy Hara Marine Protected Area and surrounding area**

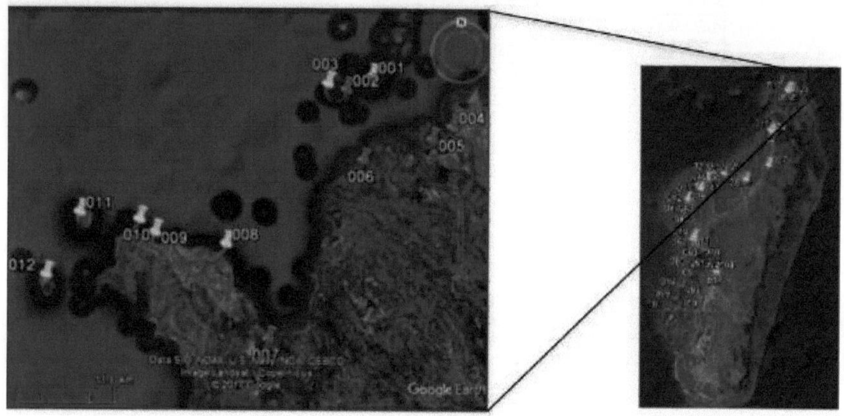

001: Nosy Nosy Hara; 002: Anjombavola; 003: Lakandava; 004: Rantavorona; 005: Andranomavo, 006: Andranoboka; 007: Anjiamaloto; 008: Fararano; 009: Andovobe; 010: Ankonkofaly; 011: Nosy Valiha; 012: Nosy Mahanonoka (Rasolonjatovo, 2018)

> **Sahamalaza-Radama Islands Marine and Coastal Protected Area and surrounding area**

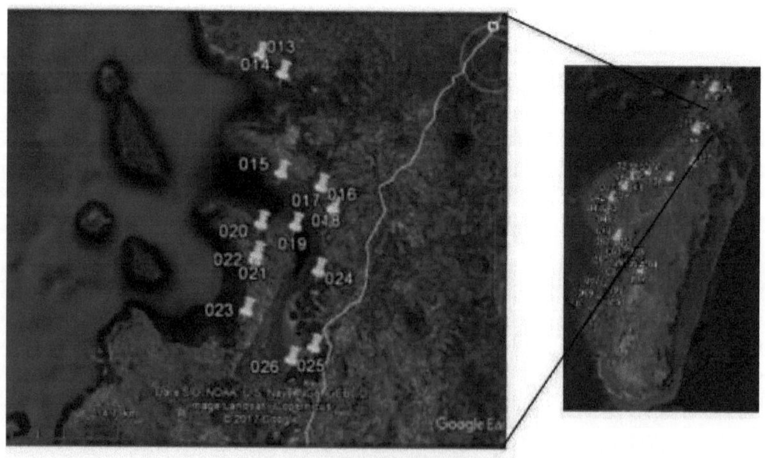

013: Andrafia; 014: Ambahofaho; 015: Nosy Ankerambalavo; 016: Anjiamalandy; 017: Vavanankaramihely; 018: Vavanambatomadoso; 019: Ankorobe; 020:Sijoro; 021: Ambohikiriny; 022: Vavanambodidimaka; 023: Tranonkira; 24: Nosy Kalanoro; 25: Ankijanimanjofo, 026: Ankitsika (Rasolonjatovo, 2018)

> **Lake Tseny and Ankarafantsika National Park**

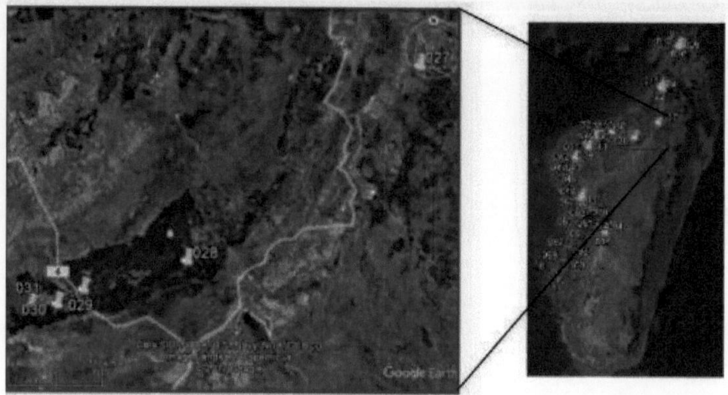

027: Tseny; 028: Tsimaloto; 029: Ravelobe; 030: Doanibe; 031: Komandria (Rasolonjatovo, 2018)

> **Baie de Baly National Park**

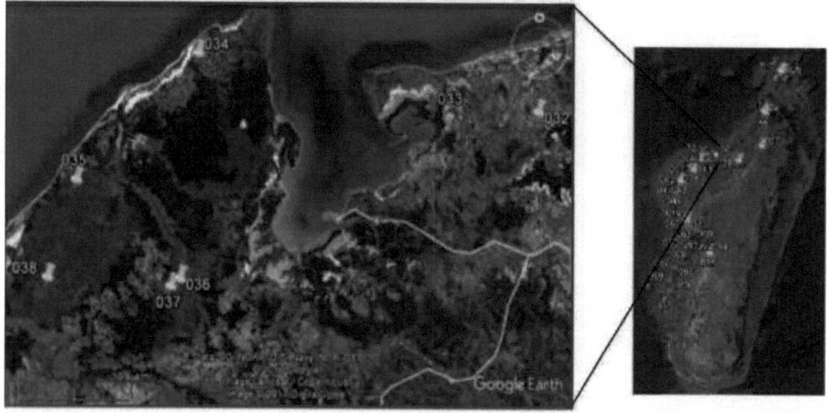

032: Ankaranajy; 033: Sanda; 034: Lac Sarka; 035: Ambatomainty; 036: Ankoro; 037: Ankoro; 038: Kapoloza (Rasolonjatovo, 2018).

> **Besalampy district**

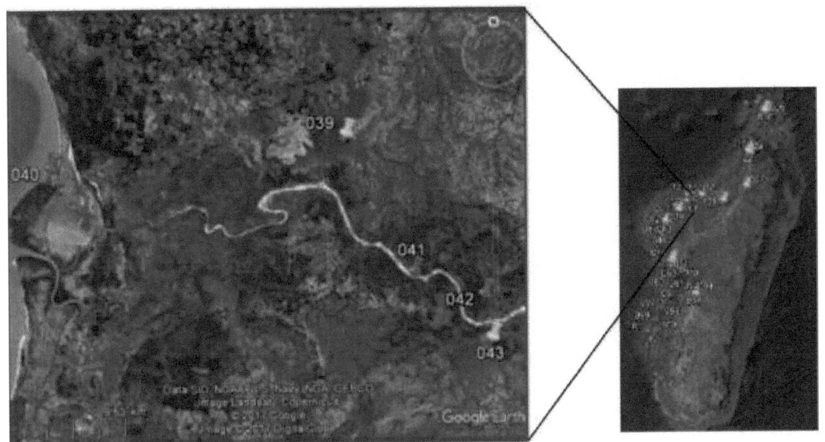

039: Sahapy; 040: Ankinga; 041: Maintimaso; 042: Ampandra; 043: Amparihy (Rasolonjatovo, 2018)

> **Maintirano district (north of the Manabao river)**

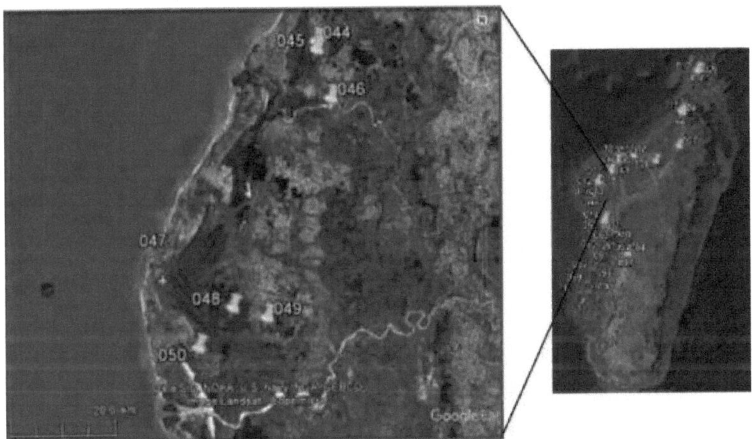

044: Bezozoro; 045: Ambalarano; 046: Andranovorinankoay; 047: Tambohorano; 048: Nosy Sarotra; 049: Mahiarere; 050: Antevamena (Rasolonjatovo, 2018).

> **Antsalova district**

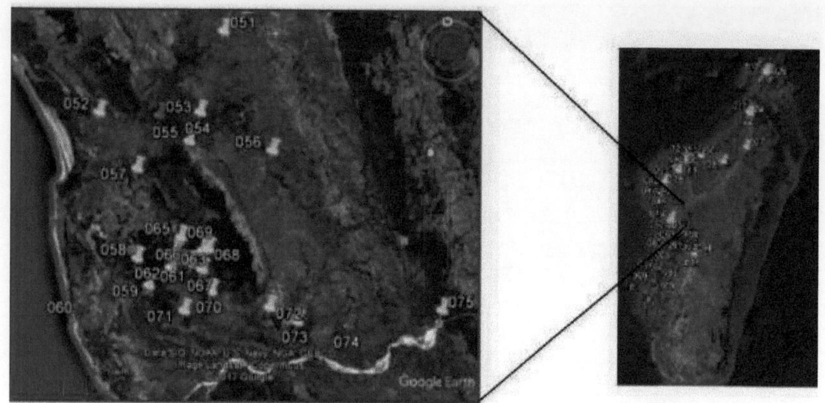

051: Androibe; 052: Soahany; 053: Ambohibary; 054: Antsahafa; 055: Masama;

056: Antsalaza; 057: Beora; 058: Andranolava-nord; 059: Antsamaky; 060: Ankelilaly; 061: Nosindambo; 062: Befotaka 2; 063: Befotaka 3; 064: Soamalipo 1; 065: Soamalipo 2; 66: Soamalipo 3; 067: Ankerika 2; 068: Ankerika 3; 069: Ankerika 4; 070: Ankerika 5; 071: Andranolava-sud; 072: Antsakotsako;073: Antsohale; 074: Ampozabe; 075: Bekopaka (Rasolonjatovo, 2018).

> **District of Belo-sur-Tsiribihina and Miandrivazo**

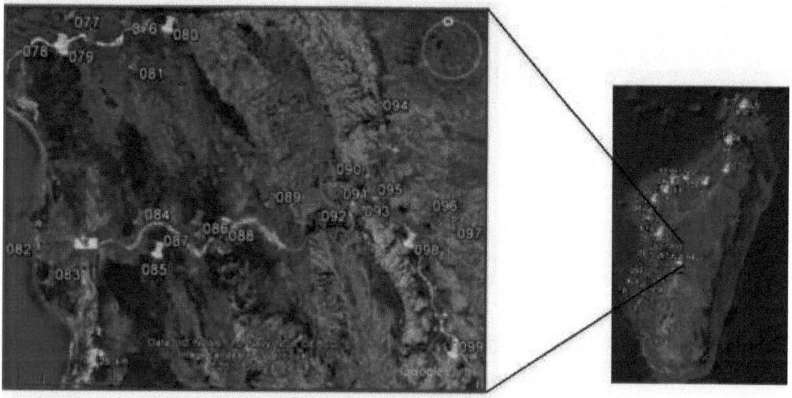

076: Tsinjorano; 077: Maromahia; 078: Ankevo; 079: Bejijo; 080: Andranovoribe;

081: Bevahomaty; 082: Soarano; 083: Nosimborona 084: Serinamo; 085: Kimanomby; 086: Akopiky; 087: Mamotsaky; 088: Ambatomainty; 089: Berevo; 090: Maroampinga; 091: Ankazomanga; 092: Anketrevo;093: Andranomazava; 094: Amparihy; 095: Borimena; 096: Tsimalaijohary; 097: Anjanaborona, 098: Asonjo; 099: Badika (Rasolonjatovo, 2018)

> **From Belo-sur-mer to Morombe**

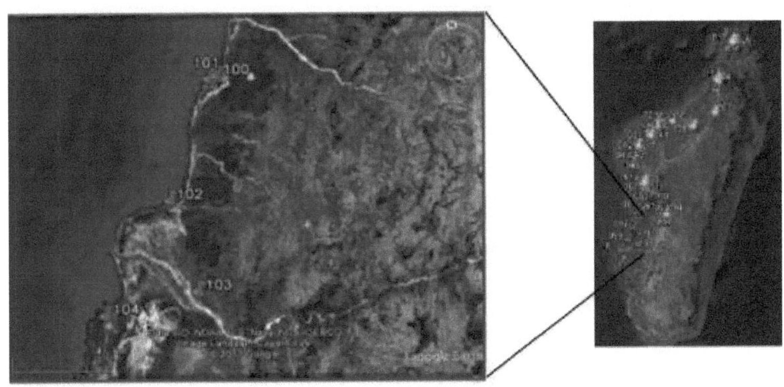

100: Morahita; 101: Amboboky; 102: Andranopasy; 103: Mangoky; 104: Morombe (Rasolonjatovo, 2018).

Legend:

, :Place not home to the Madagascar fish eagle

Location of the Madagascar fish eagle

APPENDIX 3: SOME NEST AND NEST TREE PARAMETERS

> **Continental zone (lake and river environments)**

N	Nest	Status	Nest tree	Nest height (m)	Dhp of nest tree (cm)	Distance between nest and water (m)
01	Amparihy	Former	Taly	17	77,3	226
02	Ambalarano	New	Vavaloza	16	32,2	263
03	Andranonakoay	Former	Kily	15	86	760
04	Antevamena	Former	Manga	15	46	0
05	Befotaka 2	Former	Kily	14	117,8	5,95
06	Befotaka 3	Former	Reniala	16	56,5	220
07	Somalipo 1	Former	Kily	20	91,4	18
08	Somalipo 2	Former	Reniala	18	137	126
09	Somalipo 3	New	Sarongaza	17	76,5	111
10	Ankerika 2	Former	Sarongaza	24	49,7	297
11	Ankerika 3	Former	Vory	20	82,2	20
12	Ankerika 4	Former	Sarongaza	21	62,1	292
13	Ankerika 5	Former	Anakakaraka	19	82,8	173
14	Antsahafa	Former	Adabo	15	67,6	322
15	Masama	Former	Sarongaza	16	42	40
16	Andaranolava-sud	Former	Kily	14,5	68,5	180
17	Andaranolava-nord	New	Reniala	13	170	88
18	Antsakotsako	Former	Sarongaza	16	35	45
19	Androibe	Former	Sohihy	16	92,8	2
20	Antsamaky	New	Reniala	15,5	186	802
21	Antsohale	Former	Kily	15	68	185
22	Andranovoribe	Former	Kily	14	56	220
23	Bekopaka	New	Kily	16	119	14,2
24	Antsalaza	Former	Taly	15	50	103
25	Andranomazava	Former	Sohihy	12	51,5	11
26	Asonjy	Former	Kily	17	58,1	42
	Average			**16,42**	**79,31**	**175,62**
	Standard deviation			**2,63**	**38,87**	**206,17**

> **Coastal zone**

Nest	Status	Nest tree	Nest height (m)	Dhp of nest tree (cm)	Distance between nest and water (m)
Nosy Nosy Hara	New	Hazomena	7	28,1	34
Lakandava	Former	Mantaly beravina	10	40,5	53
Ankonkofaly	Former	Farafaka	6	43,2	0
Andovobe	Former	Farafaka	7	47	0
Nosy Valiha	Former	Farafaka	5	38	0
Mahanonoka	Former	Farafaka	10	47,5	0
Fararano	Former	Farafaka	8	48,9	0
Ankij animanj ofo	New	Farafaka	10	48,2	215
Anjiamalandy	Former	Farafaka	12	38,5	0

V avanankaramihely	Former	Farafaka	13	41,6	0
Ambahofaho	Former	Farafaka	12,5	53,4	0
Andrafia	New	Farafaka	10	43,6	0
vavanambodidimaka	New	Farafaka	10	36,1	15
Ambohikiriny	Former	Farafaka	9	26,1	0
Sijoro	Former	Farafaka	inaccessible	???	0
Average			**9,25**	**41,48**	**21,13**
Standard deviation			**2,41**	**7,76**	**55,88**

APPENDIX 4: NEST DIMENSIONS

Nests	Complete structure	Height	Depth	Logis	
				Length	Width
Ankonkofaly	118	54	5	65	45
Mahanonoka	129	45	10	72	53
Befotaka III	126	63	15	74	56
Ankerika II	124	46	25	72	58
Averages ± Standard deviations	**124,25 ± 4,65**	**52 ± 8,37**	**13,75 ± 8,54**	**70,75 ± 3,95**	**53 ± 5,72**

APPENDIX 5: CHARACTERISTICS OF HABITATS SURROUNDING NESTS

> **Continental zone (lake and river environments)**

Nesting territory	Description	Distance between nest and camp (km)
Amparihy	Degraded area: where rice growers camp	0,007
Ambalarano	Dense forest dominated by Hazompasy (*Cleidion capironii*)	218
Antevamena	Degraded area near the rice growers' camp	0,053
Befotaka 1	Island dominated by Beholitsy *(Hymenoducton decarii)*	1,5
Befotaka 3	Dense forest dominated by Hazonkataky *(Rudia calcicola)*	1,7
Somalpo 1	Dense forest dominated by Katrafay (*Cedrolepsis grevei*)	1,3
Somalpo 2	Dense forest dominated by Soalafika *(Cynometra comersoni)*	0,75
Somalpo 3	Dense forest dominated by Sefo (Givotia stipuliflora)	0,64
Ankerika 2	Dense forest dominated by Nanto *(Capuronderdon perrieri)*	+ 2
Ankerika 3	Dense forest dominated by Sely (*Grewia sp*)	+ 2
Ankerika 4	Dense forest dominated by Manary *(Dalbergia* sp*)*	1,9
Ankerika 5	Forest dense dominated Anakakaraka *(Cordyla Madagascariensis)*	0,3
Antshafa	Secondary formation dominated by Mokonazy (*Ziziphus mauritani*)	0,4
Masama	Dense forest dominated by Kily *(Tamarindus indica)*	0,3
Andranolav-nord	Dense forest dominated by Hazonkataky *(Rudia calcicola)*	+ 2
Antsamaky	Dense forest dominated by Hazonkataky *(Rudia calcicola)*	+ 2
Bekopaka	Primary formation close to a cleared plot.	+ 2
Bejijo	Dense forest	+ 2
Antsalaza	Gallery forest by the river	+ 2
Andranomazava	Sacred place dominated by *Phragmites sp.*	0,7
Asonjo	Sacred island dominated by *Tamarindus induca*	1,4

> **Coastal zone**

Nesting territory	Description of nesting territory	Distance between nest and camp (km)
Nosy Nosy Hara	Rocky islet, tourist campsite and National Park guard post	0,2
Lakandava	Rocky islet, fishermen's campsite, inaccessible nesting site	0,15
Ankonkofaly	Coast with some mangroves, where fishermen camp	0,9
Andovobe	Coast with some mangroves near the village	0,6
Fararano	Islet with mangroves in the north-western part	1
Nosy Valiha	Islet with mangroves in the south-western part	+ 2
Nosy Mahanonoka	Coastline with mangroves near the village	+ 2
Ankijanimanjofo	Coastline with huge mangrove swamps	+ 2
Anjiamalandy	Coast with mangroves	+ 2
V avanakaramihely	Coast with mangroves	+ 2
Ambohikiriny	Mangrove	1,9
Vavanambodidimaka	Mangrove	+ 2
Sijoro	Coast with mangroves	+ 2
Ambahofaho	Coast with mangroves	+ 2
Andrafia	Coast with mangroves	1,7

APPENDIX 6: MADAGASCAR FISH EAGLE POPULATION IN TPF INTERVENTION SITES, BETWEEN 2006 - 2016.

^\Years Sites	2006	2007	2008	2009	2010	2011	2012	2013	2014	2015	2016
Antsalova	77	86	71	83	87	**87**	86	83	61	***55***	79
Maintirano	13	***10***	13	**20**	13	12	14	***10***	12	13	15
Total	90	96	84	**103**	100	99	100	93	73	***68***	94

Source: **TPF**

APPENDIX 7: BREEDING SUCCESS AT TPF INTERVENTION SITES

> **Successful breeding of *Haliaeetus vociferoides* in Antsalova (site 1) between 2006 and 2016.**

Year	Number of couples	Number of active nests	Number of eggs	Number of chicks	Number of chicks fledged	Breeding success rate
2016	23	19	26	15	10	62,5%
2015	17	16	28	13	13	81,25 %
2014	20	12	17	11	10	83,33 %
2013	26	21	24	18	18	85,71 %
2012	27	22	28	20	17	77,27 %
2011	27	22	30	19	19	86,36 %
2010	27	20	24	18	17	85 %
2009	26	22	29	20	17	77,27 %
2008	26	18	27	14	11	61,11 %
2007	29	24	36	19	14	58,33 %

2006	27	21	31	19	16	76,19 %
Means ± standard deviation	**24.91 (n = 274; sd = 3.64)**	**19,45 (n= 214; Sd = 3.56)**	**27.27 (n = 300; Sd = 4.80)**	**16.91 (n = 186; Sd = 3.11)**	**14.72 (n = 162; Sd = 3.29)**	**75.85% (n = 11; sd = 10.41)**

Source : **TPF**

> **Successful reproduction of *Haliaeetus vociferoides* in the wetlands of Tambohorano district of Maintirano (site 2) between 2006 and 2016.**

Year	Number of couples	Number of active nests	Number of eggs	Number of chicks	Number of chicks fledged	Breeding success rate
2016	5	4	7	3	3	75%
2015	4	3	6	3	3	100%
2014	4	3	6	3	2	66,67%
2013	5	2	4	1	1	50%
2012	6	4	7	3	2	50%
2011	5	3	6	5	2	67%
2010	5	3	7	3	1	33,33%
2009	9	3	6	4	2	66,67%
2008	5	4	7	4	1	25%
2007	5	3	6	3	3	100%
2006	6	3	5	4	1	33,33%
Averages ± Standard deviation	**5.36 (n = 59; Sd = 1.36)**	**3.18 (n = 35; Sd = 0.60)**	**6.09 (n = 67; Sd = 0.94)**	**3.27 (n = 36; sd = 1.01)**	**1.91 (n = 21; Sd = 0.83)**	**66.64% (n = 11; Sd = 25.31)**

Source: **TPF**

SUMMARY

The study of the population, distribution and ecology of the Madagascar Fish Eagle *Haliaeetus vociferoides*, an endemic and critically endangered species of raptor, was carried out in the western part of the Grande Ile, from Nosy Hara in the north to Morombe in the south, passing through Protected Areas and lake ecosystems. This study took place between July 2016 and January 2017, a period of 7 months. The count was carried out by direct observation using a pair of binoculars and a telescope. This technique was combined with individual and group surveys of the riparian population. A total of 193 individuals were recorded, divided into 68 breeding pairs, 5 solitary adults, 7 sub-adults and 34 juveniles. These pairs are made up of 57 normal pairs, i.e. pairs of two individuals, and 11 polyandrous pairs (2 males and one female). The remaining population stretches along the north-west and mid-west coast of Madagascar, specifically between Nosy Hara and Belo-sur-Tsiribihina. Maevatanana forms the inland limit. Statistical analysis showed a clear difference between the number of animals in protected and unprotected areas. The Protected Areas are home to more individuals than the other sites. There was also an increase of 1.3% in the number of individuals per year from 2006 to 2016. Finally, the Madagascar fish eagle nests on large trees with a diameter ranging from 26 to 186 cm and a height of 5 to 24 m. Hunting, the collection of eggs and eaglets, deforestation and overfishing, as well as the conversion of lakes into rice fields, are the main threats and pressures weighing on the species and its natural habitat. **Keywords** : ecology, distribution, *Haliaeetus*, Madagascar, population, *vociferoides*

Printed by Books on Demand GmbH, Norderstedt / Germany